Organic Farming and Farming Systems for Sustainable Agriculture

About the Author

Dr. S. Krishnaprabu, is graduated in B.Sc., (Agriculture) from Agricultural College and Research Institute, Madurai, Tamil Nadu, where he did his M.Sc. (Ag.) in 2002. He has been working in Annamalai University from 2002, where he obtained his Ph.D. in Agronomy also with specialization in Nutrient Management and presently he is Assistant Professor in Agronomy, his specialization is in Nutrient management and crop improvement and he has been teaching courses on Principles of Agronomy, Irrigation and dry farming management, Agronomy of field crops and Crop production to under-graduate students for ten years. He is supervising has to his credit of more than ten numbers of research papers published various international journals. Also, a widely accepted two text books entitled 'Scientific Production of Field Crops' and 'Rice Production' are value to his name.

Organic Farming & Farming Systems for Sustainable Agriculture

Dr. S. Krishnaprabu, M.Sc. (Ag.) Ph.D.
Assistant Professor
Department of Agronomy, Faculty of Agriculture
Annamalai University
Annamalainagar-608002, Tamil Nadu, India

International Books & Periodical Supply Service
(Publisher of Scientific Books)
38, NISHANT KUNJ, PITAM PURA MAIN ROAD
DELHI-110034 (India)
Phone : 011-27352078, Mobile : 09810146811, E-mail : hkjain1975@yahoo.com

Published by :

International Books & Periodical Supply Service

(Publisher of Scientific Books)
38, NISHANT KUNJ, PITAM PURA MAIN ROAD
DELHI-110034 (India)
Phone : 011-27352078, Mobile : 09810146811, E-mail : hkjain1975@yahoo.com

ISBN 978-93-88892-11-7

Composed, Designed & Printed in India

Preface

Organic agriculture is the most dynamic field and rapidly growing sector of the global food industry. Its growth from a small-scale niche to wide level enterprises has made organic and international phenomenon. Organic farming is a production system that avoids or largely excludes the use of synthetically produced agricultural inputs like; fertilizers, pesticides, growth regulators; livestock feed additives, etc. Organic farming practices are becoming more widely accepted in developed and developing countries.

Organic farming practices are eco-friendly and sustainable agricultural production system that is getting momentum with every day's rising sun.

Considering the importance of this holistic agriculture production management system most countries including India is promoting organic farming. Researchers, entrepreneurs and farmers are also very keep to work with organic farming practices.

Keeping in the view present and future prospects of organic farming this book "Organic Farming Standards, Accreditation, Certification and Inspection" is compiled. An attempt has been made through this book to cover most of the topics related to the organic agriculture to satisfy brainstorming querries to various communites. Total seven chapters are included in the book, covering all the aspects of organic production and marketing system. Some important information is given in the Annexure of the book and hope that those will be useful for the readers.

I would like to express my cordial thanks to all my well-wishers.

Author

Contents

1

Integrated Plant Nutrition System (IPNS): An Approach towards Organic Farming

The greatest challenge facing mankind in the 21st century is to produce the basic necessities of food, feed, fibre, fuel and raw materials from 0.14 ha or less land per caput. While the use of mineral fertilizers is the quickest and surest way of boosting crop production, their cost and other constraints frequently deter farmers from using them in recommended quantities and in balanced proportions. As a consequence of this and other constraints there seems to be no option but to fully exploit potential alternative sources of plant nutrients. Complementary use of available renewable sources of plant nutrients (organic/biological) along with mineral fertilizers is of great importance for the maintenance of soil productivity, *i.e.* soil structure, soil bioactivity, soil exchange capacity and water holding capacity. Results from various cropping systems and ecologies illustrate that positive interactions result from the integrated use of mineral fertilizers and organic/biological sources of plant nutrients within the framework of integrated plant nutrition system (IPNS).

In response to this need, FAO programmes on IPNS are receiving priority-an area in which FAO has been active for over a decade. This chapter provides a brief account of the concept and components of IPNS, present state of the art on the subject, initiatives and programmes of the FAO and some IPNS models which can be transferred across locations. As several of the results obtained in India have been discussed in other chapters in this volume, results included here are largely from other countries. Experimental results cited are more in the nature of examples and a detailed technical review of all available results is neither intended nor is it within the scope of this chapter.

IPNS Concept and Material

The basic concept underlying IPNS is the maintenance or adjustment of soil fertility and of plant nutrient supply to an optimum level for sustaining the desired crop productivity through optimization of the benefits from all possible Sources of plant nutrients in an integrated manner. The appropriate combination of mineral fertilizers, organic manures, crop residues, compost or N-fixing crops varies according to the system of land use and ecological, social and economic conditions. Contrary of the Low External Input (LEI) and Organic Farming approaches, the IPNS involve fertility and plant nutrition management for a targeted yield based not only on cropping and farming systems but also on distinct geographical areas or villages as a dynamic system. The IPNS approach can be modulated; a factor of targeted yields in any area according to land, water and climate potentials.

The cropping system rather than an individual crop, and the farming system rather than an individual field, are the focus of attention in this approach for developing IPNS practices for major agro-ecological zones and for various categories of farms. The best associations of various types of plant nutrients in different fields are identified for a balanced plant nutrition and high yield, at the same time sustaining soil fertility and controlling nutrient losses.

It is envisaged that locally-available materials of plant or animal origin as by-products of agricultural activities be used or, where such materials are not abundantly available, *in situ* production of organics be attempted. In, areas and farms where non-cultivated areas are available (forests, pastures and fallows), biomass production is attempted for nutrient transfer to cultivated areas in a sustainable system. In areas and farms where land is scarce and limited to the fields, increasing biomass production in fallows and on the borders of the plots can be attempted. Introduction of green manuring, of living or dead, cover in the cropping system can also contribute to nutrient supplies.

Alternative sources of organic material which have considerable potential are: quick-growing leguminous shrubs grown as a part of the cropping system and incorporated into the soil at an appropriate stage as green manure; leguminous trees grown in hedgerows and their lopping used as mulch materials or incorporated into the soil of the cropped alleys between them; forage or food legumes properly inoculated with *Rhizobium* grown in the cropping sequence; and the use of Azolla of blue-green algae with wetland crops.

The following phenomena must be understood in the context of organic re-cycling and IPNS:

1. *In situ* re-cycling of crop residues rather than their non-agricultural use and wastage will bring back a certain amount of nutrients to the same field. However, they are not new sources and are deficient in the same nutrients (like P, S, K, Ca, Mg) as those of the soil;
2. *In situ* growing of leguminous trees and their biomass incorporation will bring atmospheric N into the system, arid occasionally, through their deep-roots, nutrients leached beyond root zone and sub-soil mineralized nutrients are brought to the surface layer for the use of annual crops;
3. When biomass is brought from outside the plot or farm and/or cattle graze on uncultivated land, it is basically a transfer of nutrients from one place to another. So, except for N where biofixation can introduce nutrients into the soil-crop system, IPNS is a process which limits the losses of the system and not a process which supplies plant nutrients for the whole system.

State of the Art of IPNS

Attempts made in several countries to complement the use of mineral with organic sources of plant nutrients have generated useful, though limited information on the complementary and synergistic effects of these materials on the yield of crops. Some examples of results are described in the following sections:

Farmyard Manures (FYM)

In many countries such as India and China, cow dung is abundantly available for use as an organic source of plant nutrients. In Indian trials, the rice yield was increased by 450 kg/ha (47 per cent) by the application of 5.6 t/ha of FYM alone and by 600 kg/ha (63 per cent) when an equivalent amount of N (67 kg/ha of N) was applied as urea (Table 1.1). Applied together, the yield increase was 1120 kg/ha (118 per cent) over no N or no FYM (control plot).

In China, under low soil fertility conditions, the FYM (15 t/ha) effect on wheat yield exceeded the effect of 70 kg/ha N. When these two sources of nutrients were applied together, the yield increase over no treatment control was 460 kg/ha (35 per cent), while the yield increase with 140 kg/ha N alone was only 300 kg/ha (23 per cent).

Table 1.1: Effect of Complementary Application of FYM and Mineral Fertilizers on the Grain Yield of Rice and Wheat

Country	Crop	FYM (t/ha)	Fertilizer N (kg/ha)	Yield (kg/ha)	% Yield increasing over control
India	Rice	0	0	950	-
		5.6	0	1400	47
		0	67	1550	63
		5.6	67	2070	118
China	Wheat	0	0	1330	-
		15	0	1570	18
		0	70	1530	15
		15	70	1790	35
		0	140	1630	23
		15	140	1990	50

In another set of long-term experiments conducted at different locations in India representing varying agro-ecological situations, positive effects of FYM were observed. Thus, integrated use of organic manures with mineral fertilizers improved the soil and crop productivity and increased the yield response.

Green Manures and Crop Residues

Green manuring with legumes has long been known to be beneficial for sustainable crop productivity. With intensification of agriculture and the advent of mineral fertilizers, farmers enthusiasm for this practice declined. The lack of interest arises from farmers' perception of inadequate economic return from the labour and from the time the green manure crop occupies the land. To overcome this objection, an innovative technique is being studied in some countries, namely to plant a food legume, harvest the pods just prior to full maturity and turn the biomass into the soil, when still green.

One such attempt was made at the International Rice Research Institute (IRRI) was cowpea *(Vigna unguiculata),* of which the whole biomass was turned in at 75 per cent flowering stage and compared with the treatment where the grain was harvested before turning in the residues. Virtually no difference was found in the two years, either for upland or lowland situations, in the rice yield between the two processes. The gain from either method, averaged over the two years, was 800 kg/ ha under lowland and 500 kg/ha under upland conditions, compared with either a pre-rice fallow or a pre-rice cowpea from which all the above ground parts had been removed (Table 1.2).

Table 1.2: Grain Yield of Rice as Affected by Pre-Rice Land Use

S.No.	Pre-rice treatment	Low land (kg/ha)		Upland (kg/ha)	
		1986	1987	1987	1987
1	Fallow	4,800	4,500	1,200	1,300
2	Cowpea incorporated into soil at flowering	5,700	5,400	1,600	2,000
3	Cowpea grown, grain harvested and straw incorporated into soil	5,500	5,200	1,700	2,000
4	Cowpea grown, both grain and straw removed	4,900	4,500	1,400	1,500

Under lowland conditions, incorporating the whole biomass of cowpea at flowering, and growing rice without any fertilizer produced yields which were equivalent to applying 36 kg/ha N in 1986 and 58 kg/ha N in 1987 to the rice crop grown after a previous fallow or a pre-rice cowpea from which all the above-ground parts had been removed. Under upland conditions, these advantages were equivalent to applying 70 kg/ha N in 1986 and 65 kg/ha N in 1987 after a pre-rice fallow. The root residues of cowpea were equivalent to 30 kg/ha N in 1986 and 15 kg/ha N in 1987 under upland situations. Farmers can, thus, choose to increase yields or save on fertilizer N or a combination, depending on the economic and agronomic conditions.

Similarly, results of joint research by the Centre for Soil Research, Indonesia and IRRI during 1986/87 showed that on lowland rice incorporation of *Sesbania rostrata* at 45 days, along with an application of 60 kg/ha N enhanced rice yield by 500 kg/ha over the application of 120 kg/ha N alone (Table 1.3).

Table 1.3: Effect of N Fertilization and *Sesbania rostrata* Green Manuring (SR) on the Yield of Lowland Rice (IR-36) in East Java, Indonesia

Fertilizer kg/ha	Grain yield (kg/ha)		
	N	P_2O_5	K_{2O}
0	0	0	2000
60	45	30	3700
0 + SR at 45 days	45	30	3100
60 + SR at 45 days	45	30	4600
120	45	30	4100

While beneficial effects of mulching are widely recognized, procuring mulch material in sufficient quantity is a serious practical problem in many areas. Management of crop residue as a source of mulch is, therefore, closely linked with cropping systems, tillage methods, agro-ecological conditions, farming systems and particularly, cattle population in the farm. A range of cultural practices is available, mainly in humid tropics, to provide adequate amounts of residue mulch for soil protection and fertility enhancement, *e.g.* cover crops, conservation tillage, after cropping, use of industrial by products, etc. The sustainability of each practice depends on the local biophysical and socio-economic environment.

Legumes in Cropping Systems and Fertilizer Use

Forage legumes are well recognized as capable of improving soil fertility through symbiotic N fixation to the extent that the non-legumes subsequently grown benefit from the residual fertility. *Medicago* or annual *Trifolium* species are estimated to contribute 50-70 kg/ha N for the subsequent crop.

There is evidence that grain legumes also benefit the subsequent crop through enrichment of soil fertility. In an experiment conducted over a three-years period at Milingano (Tanzania), P applied to mungbean *(Vigna radiata)* not only increased the grain yield of the legume by 70 kg/ha (25 per cent) but also of the subsequent maize by 1000 kg/ha (59 per cent). Phosphorus application to maize via the previous mungbean crop was as effective as direct P application to maize, and the additional yield of mungbean made the system economically more attractive.

Biological Sources of Nitrogen

Biological fixation of atmospheric N by heterotrophic organisms like cyanobacteria (blue-green algae) individually or in association with the aquatic fern *Azolla* has been in use in rice production systems in many countries. Extensive trials have been made with blue-green algae and *Azolla* in several countries. In an experiment conducted in India, when mineral fertilizers were used with blue-green algae on rice, the yield was increased by 90-220 kg/ha (3 to 5 per cent over a base yield of about 4000 kg rice/ha) by their combined application.

From work done with *Azolla* in the Philippines it was observed that top-dressing *Azolla* with a basal application of 30 kg/ha N increased rice yield by 800 kg/ha (31 per cent) over mineral N alone or 400 kg/ha (13 per cent) over *Azolla* alone (Table 1.4). The *Azolla* was equivalent to about 30 kg/haN.

Table 1.4: Grain Yield of Lowland Rice as Affected by Mineral N and Incorporation of *Azolla* into the soil

Treatment	Yield (kg/ha)
T_1 30 kg/ha N	2600
T_2 *Azolla* top-dressed	3000
T_3 30 kg/ha N + *Azolla* top-dressed	3400
% increase T_2 over T_1	15.0
% increase T_3 over T_1	31.0
% increase T_3 over T_2	13.0

The importance of management of *Azolla* in the process of its incorporation into the soil and its interaction with added fertilizer phosphorus has been shown. In an experiment conducted at IRPJ, Philippines, it was noted that *Azolla* growth without P was poor. Its inoculation in the paddy field increased rice grain yield only in the plots which were not puddled. Mid-season puddling increased rice growth over shadowing the effectiveness of inoculation. *Azolla* growth was profuse in plots receiving P and the rice grain yield increased substantially with *Azolla* incorporation. An increase in rice yield of 1050 kg/ha (71 per cent) was obtained by a combination of *Azolla* inoculation, its mid-season incorporation (39 days after inoculation) and P addition over control. With only *Azolla* inoculation, but without its incorporation or P addition the yield advantage did not exceed 25 per cent. It, thus, becomes evident that proper management of *Azolla* and its incorporation in the soil are crucial factors determining its effectiveness (Table 1.5).

From this brief review presented and a wide analysis of documentation, the following conclusions can be drawn on the present state of understanding with regard to IPNS:

1. Organic/biological sources of plant nutrients complement mineral fertilizers in meeting nutrient requirements of crops. The magnitude of the contribution will vary according to sources and agro-ecological conditions.
2. In many situations, synergistic effects due to such combined application could be expected, thus increasing the fertilizer use efficiency (FUE).
3. Residual effects of added organic sources in the cropping system could also be expected together with an improvement in soil physical condition.
4. Under high input production systems, where productivity cannot be further increased with incremental use of mineral fertilizers alone, addition of organic sources could again increase yields through increased soil productivity and higher FUE.

Table 1.5: Grain Yield (kg/ha) of Rice as Affected by *Azolla* Inoculation, Method of its Incorporation and Phosphorus Addition

***Azolla* treatment**	**No P_2O5**		**35 kg/ha P_2O_5**	
	No Pudding	**Mid-season Pudding**	**No Pudding**	**Mid-season Pudding**
No inoculation	1480	2360	1860	2250
Azolla inoculation	1850	2160	2020	2530
% Difference due to *Azolla* inoculation	+25	-8	+9	+12

FAO Programmes in the Development of IPNS

FAO has taken a pioneering initiative in developing the IPNS approach within the following conceptual programmes:

Availability of Organic Sources and Potential of BNF

When planning programmes and activities in organic re-cycling, it is worthwhile, if not indispensable, to consider all the possible uses of a particular material and to examine if re-cycling it as a manure is really the best choice. Once this is clear, one can start looking into how the actually available organic materials can be re-cycled and used as manure/mulch/soil conditioner in an optimal way. Only then can sound ideas be developed on how to integrate the available nutrient carriers (organic, biological and mineral) in an environment-friendly way which is also socially and economically acceptable to the farmer.

To have a first hand appreciation of the potential availability of various sources, status reports for selected countries in the Asia and the Pacific Region and the Africa Region, covering various aspects of soil fertility and crop nutrition, were prepared. These documents show that the prospects of various organic and biological sources differ from country to country. It would, however, be necessary to quantify their actual availability for field application and, also, such assessments need to be extended to other countries. The following are the main nutrient sources, in order of priority (Table 1.6):

Table 1.6: Main Nutrients Sources

S.No.	Country	Main organic and biological sources
1	Bangladesh	Animal wastes, BNF (*Rhizobium*); green manuring
2	Indonesia	BNF (*Rhizobium*); re-cycling of legume crop residues and rice straw; animal wastes
3	Nepal	Animal wastes; BNF (*Rhizobium*); green manuring
4	Pakistan	Animal wastes; BNF (*Rhizobium*); green manuring
5	Sri Lanka	Recycling of rice straw and legume crop residues; BNF (*Rhizobium*)
6	Thailand	BNF (*Rhizobium*); crop residues; agro-industrial wastes
7	Burkina Faso	Animal wastes; crop residues; BNF (*Rhizobium*)
8	Guinea Bissau	Crop residues; BNF (*Rhizobium* and *Azolla*)
9	Madagascar	Animalwastes; crop residues, particularly rice straw; BNF (*Rhizobium* and *Azolla*)
10	Rwanda	Animal wastes; crop residues, particularly rice straw; BNF (*Rhizobium* and *Azolla*)
11	Sudan	Animal wastes; crop residues; BNF (*Rhizobium*)
12	Tanzania	BNF (*Rhizobium*); crop residues
13	Zaire	Crop residues; forest leaves; BNF (*Rhizobium*)
14	Zambia	Animal wastes (certain areas in southern, western and central provinces); crop residual; BNF (*Rhizobium*)

IPNS Models

After the identification of potential organic/biological resources, some simple models of nutrient scheduling underlying the concepts and application of IPNS for some selected cropping systems have been developed. Their details have been presented elsewhere. Attempts are being made to develop computer models for an IPNS calculation programme of plant nutrition recommendations, based on general agronomic assumptions. Farming systems should be observed and understood before proposing a new approach. Involvement of farmers in the whole innovation process is probably the most efficient way to build up sustainable and socially acceptable models adapted to different local situations.

Field Trial Network

To generate further information on combined applications of organic/ biological sources with mineral fertilizers in different agro-ecological situations and dominant cropping systems, some research institutions have been sub-contracted to initiate long-term experiments. To bypass the limitations on the number of treatments and trials that could be carried out by local research institutions in any country, and to benefit from the results of others having similar agro-ecological conditions, this field activity is being expanded into a collaborative network programme within the framework of Technical Co-operation between Developing Countries. These trials will complement the existing FAO Sponsored Fertilizer Programme in most of the countries. To start with, trials at fixed sites representing major agro-ecological conditions are being undertaken in India, Indonesia, Laos, Nepal, Pakistan, Tanzania and Thailand. In time and depending on financial resources, the network programme is proposed to be extended to other countries.

Results obtained from field experimentation can be validated against predictive models. Field experimentation can bring about a step-by-step improvement by identifying and dealing with each major constraint to plant nutrition and crop production. Moderate improvements in traditional farming may be the initial step in a long-term strategy aimed at transforming low-input subsistence farming into science-based profitable agriculture with high or moderate inputs. Thus, the above network activity is only catalytic and a beginning. Much more remains to be done to determine integrated plant nutrition and adapted practices for sustainable agriculture, according to the goals of target groups of farmers.

Promotion and Technology Transfer

Promotion of Symbiotic BNF Activities

Parallel to the field trials, the BNF technology is being developed through a number of technical co-operation projects. Major emphasis is on assisting national research institutions in the selection and production of the best *Rhizobium* strains, and in setting guidelines for maintaining high standards of inoculant quality. An expert consultation on *Rhizobium* inoculant quality was held in March 1991.

Technology Transfer

To facilitate the transfer of IPNS technology, a data bank on available results is being established.

Promising treatment combinations from the field trial network will be included in the national demonstration programme/block demonstrations within the framework of the FAO Fertilizer Programme field projects, currently operational in 21 countries.

Transfer of this technology for the benefit of small farmers.

To foster applications of technical results, FAO continues to publish publications and training materials and on appropriate methods of production, conservation and efficient use of organic/ biological sources along with efficient use of mineral fertilizers.

Future Research and Development Needs

Research Needs

It is apparent that there is a need for:

1. More information on integrated nutrient recommendations for cropping systems as a whole, taking into account balanced plant nutrition through the complementary synergistic effect of combined use of both mineral and organic/biological sources for sustained crop production and enhancing labour productivity.
2. Although general principles may be the same, technological packages for sustainable management of soil and plant nutrient resources are site-specific and depend on farming systems, farm size, availability of inputs, and socio-economic factors. Area specific and on-farm synthesis of packages is needed on the basis of components and sub-systems, such as those described above.

Agronomic productivity, economic profitability, and ecological compatibility of such packages needs to be assessed through appropriate research on well-defined, representative sites.

3. Recommendations for different agro-ecological situations, taking into account available nutrient resources, farming systems and the choice of improved agro techniques to alleviate soil constraints for plant nutrition.
4. The application of IPNS on a large scale requires some adaptive research, demonstration programmes backed by training in good management practices, and a policy improvement and infrastructure which stimulates the efficient use of both internal and off-farm inputs.

Development of National Strategy

To provide a framework for national strategies for the conservation and re-generation of soil productivity, the following activities are being initiated. These complement the ongoing activities on IPNS:

1. Selected case studies of national fertilizer and fertility situations and strategies with the following components:

   "National accounts" for plant nutrients, assessing supplies and losses and preparing balance sheets and projections by crop and homogeneous agro-ecological zone, together with identification of critical areas and total future requirements at country level.

   a) Identification and appraisal of all currently and potentially available sources of plant nutrients-organic and mineral-at country level. Review of technical and economic potentials. Assessment of options for use of these additional/alternative sources of nutrient supplies.

   b) Formulation of alternative strategies to meet future demand for soil fertility improvement and maintenance, including policy options with regard to prices, incentives, subsidies and taxes.

   c) Preparation of national action plans for the sustainable development and maintenance of soil productivity.

   A series of national programmes and projects may then follow, expanding upon experience gained through more than 25 years of activities of the FAO Fertilizer Programme.

2. Preparation of a Covenant on Good Fertilizer Use Practices.
3. Preparation of a block demonstration programme on sustainable agricultural practices, incorporating elements of sustainable development of soil productivity and crop production, such as Integrated Nutrient Management, Integrated Pest Management, Improved Soil and Water Conservation and Integrated Crop Husbandry Management, and its implementation within the framework of the ongoing block demonstration activities of the FAO Fertilizer Programme and other related field programmes.

Conclusions

Soil productivity can be improved and maintained by integrated plant nutrition systems, which address two important problems: the low and decreasing crop yields caused by plant nutrient depletion and inefficient fertilizer use in the developing world, as well as the less extensive but similarly serious problem of pollution by excessive and inappropriate plant nutrient application in developed and locally in developing countries.

The reality is that there are limitations to both low input and high input systems. Solutions will most probably have to be tailored to individual agro-ecological and socio-economic situations and are likely to involve elements of both low-input and high input systems. In most cases, minimizing external inputs through optimal combinations of plant nutrient sources available within the farm will be one of the objectives. Increased efficiency, as well as the reduction of off-farm inputs, will contribute to environmental conservation and decrease or eliminate pollution of soils and groundwater.

Each country will need to use balance sheet approaches to match the needs of its present and future populations with the potentials of its land and water resources under different input levels, and build a long-term plan of action on the results. Depending on the situation in each country, this may imply changes in policy or legislation, prices, incentives or taxes, as well as institutions.

A strategy or action plan for the sustainable development of soil productivity through IPNS would involve the mobilization of advisory services as well as of land users' Associations. It may also require technical assistance, aid-in-kind and other supply options, and revolving funds and national production of mineral fertilizers.

□□□

2

Preparation, Processing and Prevention of Organic Fertilizers

Preparation of Organic Manures

Raw Materials

Compost and farmyard manure are processed from agricultural organic wastes, such as straw, leaves, paddy husk, groundnut husk, sugarcane trash, bagasse, cattle dung and urine and habitation wastes like city garbage, nightsoil, sewage and sullage and vegetable matter. The materials undergo intensive decomposition under thermophilic and mesophilic conditions in heaps or pits with adequate moisture and finally yield a brown to dark coloured humified material in four to six months which is more stable in form, valuable for replenishments of plant nutrients, maintenance of soil organic matter and in improving the physical and microbiological conditions of the soil.

Principles of Composting

Action of Micro-organisms

The biodegradation process is carried out by different groups of heterotrophic micro-organisms, bacteria, fungi actinomycetes and protozoa. The role of cellulolytic and lignolytic microorganisms in decomposition of crop wastes and residues is of prime importance. Micro-organisms involved in the process derive their energy and carbon requirements from the decomposition of carbonaceous materials and for every 10 parts of carbon, 1 part of nitrogen is required for building up of their cell protoplasm. Fungi are more efficient in carbon assimilation than bacteria and actinomycetes. Thus carbon dioxide evolution is comparatively less when fungi are more active in biodegradation than bacterial.

When organic materials are broken down in presence of oxygen, the process is called as aerobic decomposition. Under aerobic conditions, living organisms which utilise oxygen, decompose organic matter and assimilate some of the carbon, nitrogen, phosphorus, sulphur and other nutrients for synthesis of their cell protoplasm. Heterotrophs derive energy from the decomposition of organic matter, resulting on production of carbon dioxide, humic substances and release of available plant nutrients. Carbon serves both as energy source and is also required for cell protoplasm, greater amount of carbon is assimilated than nitrogen. Generally about two-thirds of the carbon is respired/evolved as $C0_2$ and the remaining one-third is combined with nitrogen in the living cells.

Aerobic decomposition of organic materials is most common in nature and generally occurs in arable soils and in forest soil surfaces where animal droppings and organic residues are stabilised into humus, with involvement of different groups of microflora. In the aerobic process, there are no nuisance problems, such as foul odour associated with it as is produced under anaerobic conditions due to intermediate compounds.

A great deal of exothermic energy is released during the oxidation of carbon to carbon dioxide. Organic materials in compost heaps or piles under proper insulation generate substantial amount of heat which increases the temperature upto 65-70°C. However, if the temperature exceeds 65-70°C, the microbial activity is decreased due to thermal kill of the micro-organisms and the stabilization of organic matter is slowed down. Thermophilic organisms develop, when the temperature exceeds above 45°C, and they thrive best in the temperature range of 45-65°C. The major reactions likely to occur under aerobic decomposition system are as follows :

Sugars (CH_2O) x + $XO_2 \rightarrow xCO_2 + xH_2O$ + Energy
Celluloses
Hemicelluloses
Lignins
Proteins (Organic N) $\rightarrow NH_3 \rightarrow NO_2- \rightarrow NO_3-$
Organic sulphur S+ x O $\rightarrow SO_4$
Organic phosphate $\rightarrow H_3PO_4 \rightarrow$ Ca $(HPO_4)_2$
(Phytin, Lecithin)

Anaerobic microorganisms break down organic materials by a process of reduction in absence of oxygen. First a special group of acid producing bacteria, facultative heterotrophs degrade organic matter into fatty acids, aldehydes and alcohol, etc. Then a group of bacteria convert the intermediate products to methane, ammonia, carbon dioxide and hydrogen. Oxygen is also required for the anaerobic process but its source is chemical compounds and not free dissolved oxygen. Like aerobic process, the organisms use nitrogen, phosphorus and other nutrients in developing cell protoplasm. During this process the decomposition is not complete and there is less production of carbon dioxide and intermediates, like organic acid will occur in greater amounts. So, is the case of nitrogen containing substances, such as ammonia. Due to lesser microbial biomass production and carbon assimilation, there is greater production of methane. This type of fermentation takes place in gober-gas plants, waste stabilisation ponds, in marshy soils, in buried organic materials devoid of oxygen or with low oxidation-reduction potential. Intensive reduction of organic matter is also known as a putrefactive process accompanied by foul odours of hydrogen sulphide and of reduced sulphur containing organic compounds, such as mercaptans.

As compared to aerobic process where the release of energy is much greater (484-674 K Cal/glucose molecule) only about 26 K Cal of energy per gram molecule of glucose are released. The energy of carbon is in the methane gas and the resultant energy from gober-gas plants in India is utilised for cooking purpose and can also be used in running engines. The biochemical reactions that occur in anaerobic decomposition of wastes are as follows:

$(CH_2O)\,x \rightarrow x\,CH_3\,COOH$

$CH_3\,COOH \rightarrow CH_4 + CO_2$

Organic N $\rightarrow NH_3$

$2H_2\,S + CO_2 + \text{Light} \rightarrow (CH_2O)\,x\,S_2 + H_2O$

Organic P $\rightarrow$ reduced P

Factors in Process of Composting

The most important factors in decomposition of organic wastes are as follows:

1. Carbon-nitrogen ratio
2. Blending or proportioning of wastes
3. Moisture and aeration
4. Temperature
5. Reaction
6. Micro-organisms involved

7. Use of inoculants
8. Calcium phosphate, and
9. Destruction of pathogenic organisms.

C:N Ratio

The carbon-nitrogen ratio of organic materials is the most important aspect of composting. The process of conversion of organic materials into manure is chiefly microbiological and is, therefore, influenced by the proportions of carbonaceous and nitrogenous materials that are present in organic wastes to start with. Micro-organisms need carbon for growth and nitrogen for protein synthesis. It wag found that a C: N ratio of 30 of raw materials could be most desirable for efficient composting. The C: N ratio between 26 to 40 as reported by many workers, provide for rapid and efficient composting. If the organic material is poor in nitrogen or in other words, carbon nitrogen ratio is wide, biological activity diminishes and several successions/cycles of organisms may be required to degrade carbonaceous materials. Immobilised nitrogen is re-cycled on the death of some of the organisms. Thus, the limited nitrogen is recycled by reducing the carbon content of the organic wastes. At low C:N ratio, ammonia is formed which under favourable conditions can be further oxidised to nitrite and nitrate. Other nutrients, like phosphorus, potassium, sulphur and micronutrients are also essential for microbial growth.

Lower C:N ratios under unfavourable conditions cause increasing loss of ammonia and higher values may prolong the period of composting. High C:N ratios are generally caused by organic materials poor in nitrogen such as cereal residues of wheat, paddy, *jozvar, bajra,* maize, sugarcane, trash, stalks of cotton, jute and sawdust. Farm and urban wastes have been found to vary widely in C : N ratio ranging between 30 and 80. As the stock of available nitrogen gets exhausted, the activity of nitrogen fixing micro-organisms pre-dominates and thus, there is a gain of nitrogen fixed from the atmosphere. With a C: N ratio of raw materials less than 30:1, the proportion of nitrogen is in excess of the requirements of micro-organisms. Although the process of decomposition goes on uninterrupted the unassimilable nitrogen is lost in the form of ammonia gas.

Similarly, 20 to 40 per cent nitrogen is lost in the form of ammonia during preparation of organic manure from cattle-dung and night-soil. This implies that a thorough mixing of these fractions with the carbonaceous wastes cannot be attained in practice and there are always pockets of concentration of nitrogenous materials which cause an imbalance in the decomposition process and loss of nitrogen. It is obvious that simple nitrogenous compounds are more susceptible to loss which are easily decomposed by micro-organisms and these compounds are of high utility to plants. Control of the loss and retention of this important plant nutrient in the final manure would be of practical value in obtaining a better quality manure.

The time of composting can be reduced by adding a nitrogen source of organic materials, such as activated sewage sludge or by blending with organic residues richer in nitrogen, such as legume residues and aquatic weeds (water hyacinth), slaughter-house wastes, green leaves, wastes of sugar, and wastes of antibiotic, yeast and paper industries.

Shredding and Blending

Composting process can be accelerated, if raw materials are shredded into smaller pieces or ground since the material becomes more susceptible to bacterial attack due to exposure of greater surface area. The most desirable particle size for composting is less than 5 cm although larger sizes can be composted satisfactorily. However, the advantages of shredding may be uneconomical while composting on farms and in villages or by individual farmers.

The C/N ratio and moisture percentage are the two important parameters which should be considered while mixing or blending different type of organic wastes for composting. There may not be any need of blending if the C/N ratio of organic materials is between 25 to 50, although 30 to 40 may be a good range. Substances poor in nitrogen such as saw dust, paper, straw, etc. can be mixed with comparatively nitrogen rich materials, *e.g.* legume residues, water hyacinth, slaughter-house wastes, fish scrap, night-soil, sewage sludge and biogas spent slurry to obtain a near optimum C/N ratio between 30 to 40. Similarly materials too dry and too moist can be blended together to obtain a desirable moisture level to avoid anaerobic conditions leading to bad odour.

Soil at 5 to 10 per cent is also sometimes added to compost to reduce the moisture content and to absorb ammonia in low C/N ratio materials. The dry soil may be added if sufficient dry materials are not available. Soils may be added to high C/N ratio organic matter to buffer acid conditions and to act as a diluent for retarding anaerobic process.

It can also be used for improving the appearance of finished compost to give it more granular structure.

Moisture

If the amount of moisture in the compost pile is below 40 per cent (W/W), decomposition will be aerobic but slow. Aerobic decomposition can go on at any moisture content between 30 and 100 per cent, if adequate turnings are provided. During aerobic composting high moisture content should be avoided so that anaerobic conditions are not created. The optimum moisture level of 50 to 60 per cent may be quite satisfactory for aerobic composting. However, it may be in range of 40 to 80 per cent depending on the nature of organic material to be composted. At IARI, when composting fibrous materials containing straw were composted aerobically. 80 to 85 per cent moisture was maintained. If anaerobic composting is to be carried out, the maximum moisture content is not important. The moisture content between 80 to 90 per cent may be maintained.

Temperature

With the multiplication of micro-organisms in the composting mass, the heat of exothermic biological reaction is retained due to sufficiently large mass and when the temperature rises above 40°C, the mesophilic micro-organisms are replaced by thermophilic. High temperatures are essential for destruction of pathogenic organisms and weed seeds. This generally occurs within 2 to 5 days of the start of the composting. The temperature in the middle of the pile goes up to 55° to 70°C and after this it gradually cools to ambient temperature. Decomposition is fastest in the thermophilic stage. The optimum temperature based on oxidation of organic matter into carbon dioxide and water has been found to be 60°C. Schulz showed that a maximum temperature of 71°C was optimal, but it was found that temperature should not exceed 70°C for long because decomposition will be slowed down by a thermal kill of micro-organisms and only a few of the thermophilic organisms actively carry on decomposition above 70°C. However, while composting high C/N ratio materials, such as wheat and paddy straw, *joxvar* stalk and *jamun* leaf fall, etc. of C/N ratio 48 to 50 in cubic metre pits, it was observed by Gaur that temperature of the compost pile did not go beyond 52°C. This indicated that extent of rise of temperature in compost will depend on the type material being composted and probably also on the size of the heap.

Reaction

The initial pH in compost heaps is generally slightly acidic round pH 6 as is found in the cell sap of most of the plants. The production of organic acids during the early stages of composting causes further acidification (pH 4.5-5.0) but as the temperature rises, the pH increases to slightly alkaline (pH 7.5-8.5) reaction.

Microbiology

The microbial population changes during aerobic composting, Chang has described a typical pattern. The fungi and acid producing bacteria appear during the initial mesophilic stage. As the temperature increases above 40°C, these are replaced by thermophilic bacteria, actinomycetes and fungi. Spore forming bacteria develop at temperature above 70°C. Finally, mesophilic bacteria and fungi re-appear as the temperature falls down.

Many aerobic mesophilic bacteria initially present in the composting material multiply and show increased activity. As the temperature is raised, their numbers decrease due to change in environment. A minimum is reached at 55° to 65°C during composting of grass cuttings and straw but their number increases again as temperature drops below 45° to 50°C.

The role of bacteria may be involved in raising the temperature of the compost for the development of thermophilic micro-organisms which colonise. Mesophilic bacteria which flourish during limited time, consume the most readily degradable carbohydrates. During the process compounds rich in nitrogen, such as proteins, amino acids, blood-meal and peptones are bio-degraded especially by spore forming *Bacillus* spp. Actinomycetes degrade starch actively and also bring about large losses of water soluble fractions. Thermophilic bacteria do not appear to be important in degradation of cellulose and lignin but attack protein, lipids and hemicellulose fractions. Lipids are degraded to a great extent whereas degradation or celluloses and hemicelluloses is comparatively slow and intermediate. Lignin are the most resistant to decomposition and persist for longer period and tend to accumulate.

Actinomycetes (*Thermomonospora curvata*) may be important in cellulose decomposition. *T. curvata* was the most frequently occurring actinomycete in municipal and mushroom composts, because therrmophilic actinomycetes can grow at higher temperatures than thermophilic fungi and they become dominant at the warmest stage.

Mesophilic fungi are present as the compost warms up moderately. These fungi are saprophytic sugar fungi which are quickly replaced by thermophilic fungi. The mesophilic fungi re-appear in large numbers as the heap cools down below 40°C. Evidently they persist in outer layers during the thermophilic stage and re-appear when the temperature drops. They can utilise cellulose and hemicellulose but not as efficiently as thermophilic fungi.

Thermophilic fungi occur, when the temperature ranges between 40°C and 60 C. Like mesophilic fungi, they survive in the periphery of the heap when the temperature exceeds this range. The bacteria, fungi and actinomycetes isolated and reported from compost are presented in Table 2.1.

Table 2.1: Bacteria, Actinomycets and Fungi Isolated from Compost

BACTERIA	FUNGI
Mesophilic	Mesophilic
Cellumonas folia	*Fusarium culmorum*
Myxococcus virescens	*F. roseum*
M. flavus	*Coprinus cinerus*
Thiobacillus thiooxidons	*C. megacephalus*
T. denitrificans	*C. lagopus*
Aerobacter spp.	*Clitopilus pinsitus*
Proteus spp.	*Aspergillus niger*
Pseudomonas spp.	*A. terreus*
Thermophilic	*Geotrichum candidum*
Bacillus stearothemophilis	*Rhizopus nigricans*
ACTINOMYCETES	*Trichoderma viride*
Thermotolerant and Thermophilic	*T. (lignorum) harzianum*
Micromonospora vulgaris	*Oospora variabilis*
Nocardia brasiliensis	*Mucor spinescens*
Pseudonocardia thermophila	*M. abundans*
Streptomyces rectus	*M. variens*
S. thermofuscus	*Cephalosporium acremonium*
S. thermoviolaceus	*Chaetomium globosum*
S. thermophilus	*Glomerularia* sp.
S. violaceoruber	*Fusidium* spp.
Thermomonospora curvata	*Actinomucor corymbosus*
T. fusea	*Mucor jansseni*

T. glaucus	*Talaromyces (Penicillum) variabile*
Thermopolyspora polyspora	*Helminthosporium sativum*
	Aspergillus fumigatus
Thermotolerant and Thermophilic	
Humicola insolens	*S. chiorinum*
H. griseus var. *thermoidesus*	*C. t. 6 (Mycelia sterilia)*
H. lanuginose (Thermomyces lanuginosus)	*Stilbella thermophila*
Mucor pusillus	*Malbranchea pulchella* var. *sulfurea*
Chaetomium thermophile	*(Thermodium sulfureum)*
Absidia ramose	*Dactylomyces crustaceous*
Talaromyces (Penicillium) duponti	*(Thermoascus aurantiacus)*
T. thermophilus	*Byssochlamys* spp.
T. emersonii	*Torula thermophila*

Inoculants and Other Preparations to Compost

If C/N-ratio is too wide, adding nitrogen can hasten composting by lowering the ratio. Microbial inoculations may also be beneficial in a compost heap poor in micro-organisms for speeding up decomposition of materials rich in cellulose and lignin. It has been observed in sawdust compost that inoculation with spores of cellulolytic fungus, *Coprinns ephemerus* and addition of nitrogen, phosphorus and potash shortened composting of sawdust from usual 1-2 years to 3 months and produced a compost that did not immobilise nitrogen as compared to fresh sawdust.

Japanese patents for additives to accelerate decomposition of wood products have been reported recently. The addition of nitrohumic acid (humus or complex natural ligno-protein treated with nitric acid) to bark wastes and chicken manure (1:7:100) reduced the decomposition time from 180 days to 60 days. Decomposition of wood pulp was improved by urea and crude acetic acid distilled from wood. The recent work done in India on microbial inoculants for accelerating the compost-making has been discussed separately in this chapter.

Calcium Phosphates

Several investigators have reported that calcium phosphate increases the rate of decomposition and nitrogen conservation, but more than 2 per cent inhibited the decomposition rate. Chang found increased rate of decomposition occurring mainly in cellulose and hemicellulose fractions. The number of cellulolytic microorganisms increased with the addition of phosphate and conserved nitrogen by decreasing the number of denitrifying bacteria. Although N fixation may be increased by adding calcium phosphate, there is generally sufficient phosphate in the organic wastes for meeting the requirements of micro-organisms which need about 5 to 20 per cent only as much P as N. However, our experiments have shown that application of rockphosphate is important for nitrogen conservation and enrichment of organic manures, while composting wide C/N ratio organic materials (paddy, *jowar* straw, *jamim* leaves). Gaur recorded that *Azotobacter* numbers in compost were greater in presence of added rock phosphate.

Gaur reported that addition of rockphosphate at 1.0 per cent (W/W) increased the decomposition of paddy straw by cellulolytic fungi (*Aspergillus* spp. supplemented with urea nitrogen) (Table 2.2). The carbon content was decreased by 17.2 per cent and nitrogen augmented by 31.4 per cent. As a result C/N ratio of 23.1 was obtained within 8 weeks as compared to 37.7 in the control. Increased amounts of rockphosphate (2 and 3 per cent) showed slight reduction in decomposition intensity. Mathur from Ranchi Agricultural College

Table 2.2: Effect of Rock Phosphate on the Decomposition of Paddy Straw by *Aspergillus* sp.

Treatments	% Loss in Weight (2 Months)	Organic Matter %	Carbon %	Total N%	C/N Ratio
Urea (0.5 per cent w/w) control	45.2	75.35	43.71	1.190	37.13
Urea + Rock phosphate (1.0 per cent w/w)	55.0	62.36	36,17 (-17.2)	1.564 (31.4)	23.12
Urea + Rock phosphate (2.0 per cent w/w)	52.9	62.64	36.33 (-16.9)	1.326 (11.24)	27.40
Urea + Rock phosphate (3.0 per cent w/w)	52.8	64.60	37.47 (-14.3)	1.275 (7.1)	29.39

Figures in parentheses show per cent loss or gain.

reported that quality of compost prepared from mixture of paddy straw, grass and water hyacinth was improved when rock phosphate was applied to it with and without pyrite. The chemical composition of the compost is given in Table 2.3. The results showed that low dose of rock phosphate (5 per cent) increased the nitrogen content by 30 per cent of the compost over the control. The enrichment of compost with rock phosphate besides increasing the phosphorus content, increased the concentration of calcium and micronutrients particularly of iron, manganese and zinc. Beneficial effect of addition of rock phosphate at an optimum rate during the initial stage of compost making has been reported at IARI and IC AR Co-ordinated Project on "Microbial Decomposition and Re-cycling of Organic Wastes."

Table 2.3: Chemical Characteristics of Compost Prepared with Rock Phosphate

Components	T_1	T_2	T_3
Total Ca (%)	1.85	2.27	1.92
Total Mg (%)	0.37	0.51	0.38
Total Na (%)	0.06	0.19	0.09
Total K (%)	0.92	0.81	0.83
Fe (ppm)	5,080	12,720	28,200
Mn (ppm)	180	520	540
Zn (ppm)	60	95	80
Cu (ppm)	10	24	44
pH	4.40	5.55	5.80
Organic Carbon (%)	14.64	18.06	12.83
Total Nitrogen (%)	0.61	0.19	0.58
C/N ratio	24.00	22.80	22.10

T_1: 650 kg raw material (P_0)
T_2: 650 kg raw material + 32.5 kg Mussoorie rock phosphate (MRP)
T_3: 650 kg raw material + 32.5 kg (MRP) + 32.5 kg Amjhore Pyrite.

Destruction of Pathogens and Parasites

The destruction of pathogens and parasites of animals and humans is an important aspect and a problem when compost is prepared from materials containing sewage sludge and night soil. Aerobic composting at high temperatures is effective in killing pathogens. Gotaas recommended 60°C for Thermal Kill of common pathogens listed in the Table 2.4.

Table 2.4: Temperature and Time of Exposure Required for Destruction of Some Common Pathogens and Parasites

Organism	Observations
Salmonella typhosa	No growth beyond 46"C; death within 30 minutes at 55°-60°"C and within 20 minutes at 60°C; destroyed in a short time in compost environment
Salmonella spp.	Death within 1 hour at 55°C and within 15-20 minutes at 60°C
Shigella spp.	Death within 1 hour at 55°C
Eschirchia coli	Most die within 1 hour at 55°C and within 15-20 minutes at 60°C
Enlamoeba histolytica cysts	Death within a few minutes at 45°C and within a few seconds at 55°C
Taenia saginata	Death within a few minutes at 55°C
Trichinella spiralis larvae	Quickly killed at 55"C; instantly killed at 60°C
Brucella abortus or *Br. suis*	Death within 3 minutes at 62°-63°C and within 1 hour at 55°C
Micrococcus pyogenes var. *aureus*	Death within 10 minutes at 50°C
Streptococcus pyogenes	Death within 10 minutes at 54°C
Mycobacterium tuberculosis var. *hominis*	Death within 15-20 minutes at 66°C or after momentary heating at 67°C
Corynebacterium diptheriae	Death within 45 minutes at 55°C
Necator americanus	Death within 50 minutes at 45°C
Ascaris alumbricoides eggs	Death in less than 1 hour at temperatures over 50°C

The extent and duration of high temperatures in compost heaps as well as antibiosis as a result of mixed population of micro-organisms provide reasons to believe that pathogens and parasites are not able to survive the composting process. To achieve this at least 2 to 3 turnings of compost heap is suggested. Knoll found that 65°C for one day ensured elimination of *Salmonella* spp. Stranch reported the destruction of *S. enteriditis, Erysipelothrix, Rhasiopathiae* and the *Psittacosis* virus under similar conditions. *Bacillus anthracis* was destroyed only at temperature above 55°C for 3 weeks or longer at a moisture content of 40 per cent or more. He concluded that higher temperature may not be the only way to kill pathogens during composting but some are killed by competition with other micro-organisms. Schraff (1940) and Scott (1953) found that resistant forms of parasites, such as *Ascaris* eggs, cysts of *Entamoeba histolytica* and hookworm eggs were destroyed during aerobic composting. Scott found that at least 2 turns were necessary for the destruction of *Entamoeba histolytica* in 12 days and three turns were required to destroy the eggs of *A. lumbricoides* in 36 days. Anaerobic composting at the mesophilic temperature range does not effect satisfactory destruction of parasites in a relatively short time. Anaerobic composting when using sewage sludge or night soil should be preceded by aerobic composting at least for a week. The natural death of pathogens and parasites will occur in an anaerobic environment and the microbial antagonism will eventually eliminate them but this will generally take longer period of about six months.

Heaps versus Pits

When preparation of manure is carried out in over-ground heaps, the decomposition becomes mainly aerobic, with air freely accessible for micro-organisms to carry out their activities. The rate of decomposition is accelerated, if periodical turnings are given which also promotes physical disintegration of the raw materials. Owing to quick oxidation of organic matter, high temperatures (60-70°C) develop, thereby killing weed seeds and pathogenic organisms. Loss of organic matter in this process often amounts to above 50 to 60 per cent and decomposition may continue even after the C/N ratio is reduced to the optimum. Loss of nitrogen from 20 to 50 per cent invariably takes place in this method.

Moisture content necessary for microbial activity is effected in the heap method, with losses occurring because of exposure to the atmosphere and the high temperatures that develop in the manure heap. In dry climate, this would affect the rate of decomposition and necessitate addition of extra water periodically. During the rainy season, on the other hand, there may be too much of water in the heap to slow down decomposition and even cause loss of nutrients and organic matter by leaching, unless manure preparation is carried out under sheds. This is not always feasible where large quantities of manure are required to be prepared.

Preparation of manure in pits has several advantages over method of preparation in heaps over ground although it involves some initial expenditure on digging and permanent occupation of land for this purpose. The decomposition is mainly anaerobic, except in the surface few inches and proceeds slowly. Losses of organic matter as well as nitrogen are consequently reduced, amounting to about 25 and 20 per cent respectively, enabling a larger quantity as well as good quality of the manure to be obtained. However, the process of composting can be accelerated by providing a few turnings at periodic intervals which will create aerobic conditions as well. The materials are protected against loss of moisture by evaporation as well as excess of water in the rainy season and physical loss of nutrients by leaching. High temperatures do not develop in this case owing to the anaerobic nature and the slow rate of decomposition. But weed, seeds and pathogenic organisms are destroyed because of the toxic products of this type of decomposition. Little physical disintegration of material, such as straw and refuse, however, takes place in this system but this is not necessary from scientific standards of assessment of quality of manure of which a final C/N ratio of less than 20:1 indicates its maturity and fitness for application to land. The manure may remain stored in pits until required without much loss due to very restricted decomposition after it is ready.

Utilisation of Urine Fraction

The liquid excreta of animals contains nitrogen in the form of urea, phosphorus, potash and several other nutrients in water soluble form which are of high utility to plants. Stress has always been laid on proper collection of this fraction and its utilization by adding urine-soaked litters or urine-earth to the manure-pit. However, in the process of its microbial fermentation with carbonaceous materials, the original simple compounds of nitrogen are converted into complex microbial proteins or into ligno-proteins by chemical reaction. There is also considerable loss of nitrogen in the form of ammonia due to the ease of decomposition of urea by numerous micro-organisms. An overall reduction in the manurial value of this fraction thus takes place when it is incorporated in manure heaps. The low C/N ratio of urine and the simple nature of its nitrogenous compounds suggests that its separate collection and direct utilisation as manure would be a more effective alternative, if practical methods of collection and conservation of this liquid manure are evolved. Work on this aspect carried out at the Indian Agricultural Research Institute and National Dairy Research Institute showed that urine-soaked earth can be treated with superphosphate or gypsum to control the decomposition of nitrogen and then quickly dried to stop bacterial activity or urine-soaked ground paddy husk or saw-dust can be composted along with dung with a good quality manure. The plant nutrient constituents, thus, remain conserved in their original form and the manure obtained is highly effective. Alternative materials like dry leaves, dry dung, etc. can be used in conjuction with soil in cattle-sheds to soak urine more effectively and processed as above. *In pucca* cattle-sheds where urine and cattle-shed washings are collected in cess pits, the liquid, manure may be similarly treated with superphosphate or a mineral acid cake hydrochloric acid and then soaked and dried in absorbing materials of the type mentioned.

Results obtained at the Indian Agricultural Research Institute have shown that the method can be effectively extended and applied to the collection and utilisation of large quantities of human urine which has not been exploited so far. A simple design of urinals evolved by the Institute (IARI) enables urine to be collected in absorbing materials contained in a drum or a pit dug for the purpose. The decomposition of urine is controlled by acidulating the material with superphosphate or hydrochloric acid. The material can be repeatedly soaked and dried to obtain concentrated manures. Dehydrated human urine manures have been found to be highly effective in stimulating growth and yield of crops. The proposition of separate collection and utilisation of cattle and human urine in the above manner offers a new yet unexplored field for efficient utilisation of an inexhaustible and rich source of manure available in the country.

Table 2.5: Effect of Human Urine Dehydrated in Different Materials on Wheat (Manures Applied at 112 kg/ha)

Treatment	Yield in g from 30 plants		
	Grain	Straw	Total
No manure	11.5	23.8	35.3
Urine in soil	26.6	33.9	60.5
Urine in saw-dust	27.4	37.5	64.9
Urine in charcoal dust	28.0	37.3	65.3
Ammonium sulphate	22.8	51.6	54.4
C.D.	5.7	4.8	7.6

Preparation of Farmyard Manure

Organic waste materials that generally go in production of farmyard manure including dung, urine, soaked earth or litter, remnants of cattle food, ash and other household and kitchen wastes. The existing method of preparation by dumping the small daily collection in a heap or shallow pit leaves much to be desired and yields a low-quality manure.

Sectional Filling

One of the effective methods of preparation of farmyard manure that has been found to yield superior quality manure is the process called 'Sectional Filling of Trenches', which includes all the advantages of carrying out of decomposition in pits and suits well to the conditions existing in this country of having to deal with a comparatively small daily collections. The method consists of digging suitable-sized trenches depending on the number of cattle as shown in Table 2.6.

Table 2.6: Methods of Digging Trenches

No. of cattle	Length (m)	Breadth (m)	Depth (m)
2 to 5	6.5	1.0	1.0
6 to 10	8.0	1.2	1.0
11 to 20	10.0	1.4	1.0
Above 20	10.0	1.6	1.0

Filling of mixed wastes from the cattle-shed is recommended from one end of the trench in a section, one metre in length. A movable partition made from cotton stalks or such other material between two bamboo poles is used for making the section to be filled. The daily collection is dumped in this and covered with a thin layer of soil. This is continued until the section is filled to a level of about 50 cm above ground level. The top is then covered with earth and the partition is moved one metre further along the length to form the next section for filling as before. Leaves and other farm wastes like paddy and groundnut-husks, straw, etc. could be added to increase the quantity of manure, if sufficient litter from the cattle-shed is not available. The manure is ready in about three months and contains about 2 per cent nitrogen on dry weight basis. The quantity of manure obtained per pair of animals amounts to 10 to 12 tonnes per year.

Composts

The principles of production of farmyard manure from cattle-shed wastes have been extended for the preparation of 'composts' in which a variety of carbonaceous and nitrogenous waste materials may be utilised. Thus, farm-wastes like straw, chaff, leaves, paddy and groundnut-husks, sugarcane-trash, bagasse and other agricultural as well as industrial and habitation wastes of all sorts like street-refuse, night-soil, slaughter-house wastes, sewage and sullage water, sludge, etc. can be pressed into service to yield manures.

In India, pioneering work in this field was carried out by Howard at Indore and Fowler at Bangalore, who standardized conditions for degradation of leaves, straw and town-refuse, utilizing dung and nightsoil as starters for promoting bacterial activity.

Indore Method of Composting

In this method the waste materials are spread out the moistened with dung or night-soil slurry. These are then built up into heaps, four to six metres in length, one metre in breadth and one metre in height. Under these conditions, the rate of decomposition is rapid and high temperatures develop quickly. The process is accelerated if turnings are given, whereby aeration and mixing of the materials and moistening at this stage, if necessary, cause more or less total disintegration, yielding a brown homogeneous manure in about three months. Under this aerobic process of decomposition losses of organic matter and nitrogen are heavy, amounting to 40 to 50 per cent of the initial. The manure resembles the traditional farmyard manure in appearance and properties and it has been shown that decomposable organic materials of any type can thus be processed to supplement the supplies of manure. This method of manure preparation, however, involves considerable labour in building up the heap to proper shape and periodical turnings that are given and becomes expensive and impracticable where large quantities of materials have to be processed. However, the turnings of the mass are not necessary and decomposition can go on to the desired extent, if adequate moisture level is maintained. The average composition of manure prepared by the Indore method has been found to be 0.8 per cent nitrogen, 0.3 per cent phosphoric acid and 1.5 per cent potash.

Fowler worked out the process of 'activated compost' in which fresh raw materials are incorporated in an already fermenting heap so that the established and large microbial population brings about quicker decomposition The process is useful particularly where offensive materials like night-soil are to be quickly and effectively disposed off.

Compost from Habitation Wastes-Bangalore Process

Pioneering work in preparation of manure in pits has been carried out in India by Acharya particularly on the utilisation of town-refuse and night-soil. What is called the hot fermentation method of manure production or the Bangalore process has by now been adopted in many towns in India, solving effectively one of the most difficult problems of sanitary disposal of these offensive wastes and yielding a high quality organic compost manure in the bargain. The operative part of the process can be adapted to the prevailing system of collecting refuse and night-soil. The method carried out in practice is as follows:

The compost production depot is located just on the outskirts for convenient transport of street-refuse and night-soil to the pits. In large towns, mere may be two or three such depots to serve different localities. The compost depot should accommodate about 200 trenches of dimension as under, which can take the dumpings for one year (Table 2.7).

Table 2.7: Trenches Dimensions

No. of cattle	Length (m)	Breadth (m)	Depth (m)
Below 10	4.5	1.5	1.0
10 to 20	6.0	2.0	1.0
20 to 50	9.0	2.0	1.0
Above 50	10.0	2.5	1.0

The trenches are dug in rows, with a space of 1 to 5 m between trenches. Roads of suitable width are provided between rows for the carts to approach and unload the materials inside the trenches. At first the refuse is dumped into the trench and spread out with rakes to make a layer of 15 cm. Night-soil is then discharged over this and, if necessary, spread out with long wooden spades in a layer of about 5 cm. This is then covered with a 15 cm layer of refuse. Night-soil and refuse, thus, follow in alternate layers until the pit is filled to 15 cm above ground level, with a final layer of refuse on the top. This may be dome-shaped and covered with a thin layer of soil In some towns, night-soil comes mixed up with refuse in which case the material is simply dumped to fill the trenches. Sullage water, if collected in carts as in some towns, may also be emptied over the layers of refuse. In fact, the system provides a method of disposal of every kind of waste, that the town may have, including slaughter-house, carcasses of animals, sewage, etc.

The decomposition of dumped materials in pits takes place largely in the absence of sufficient air except in the surface foot or so. This anaerobic decomposition is comparatively slow but markedly less wasteful. The difference in recovery of organic matter and nitrogen by this method as compared with the aerobic method in over-ground heaps is given below for manure prepared from town-refuse and night-soil.

Table 2.8: Percentage Recovery

	Heap	Trench
Organic matter	53	78
Nitrogen	54	74

In this method, no high temperatures develop in the lower layers and since the material does not receive turnings, disintegration into a homogeneous mass of manure does not take place. Even so, the C/N ratio is reduced to less than 20 : 1 in about six months and the manure is then ready for use. Deeper in the trenches, the material still possesses an unpleasant smell and it is desirable to dig it out and pile it up outside for some days before disposal.

In the absence of facilities for digging trenches, town-refuse and night-soil compost can be prepared in over ground heaps of one metre height, one metre width and any convenient length, with refuse and night-soil, placed in alternate layers, as in trenches and a final refuse layer on the top. Night-soil carts empty the material along the length of heap. No turnings need be given to the heaps and the material decomposes more quickly than in pits and can be used after three to four months.

The basic aspect of these processing treatments in manure production is that after the initial active decomposition and loss of excessive carbon has taken place, the proportion of nitrogenous night-soil and organic carbonaceous material is just proper and it can be used as manure. Often, the night-soil and refuse layers can be identified, when the manure is dug out. The material gets mixed up only while it is being taken out, but because the mixture is in proper proportion, it is safe to apply as manure. Further, decomposition is carried out in the soil itself and because of the presence of adequate nitrogen this causes no adverse effect and instead soon releases nitrogen for crops to assimilate.

In villages individual latrines are limited to only a few rich families, hence night-soil collection and its compost making is riot popular. It can, therefore, be a great step in improving the existing in sanitary conditions to provide suitable community or individual laterines, preferably Wardha-type trench latrines to villagers to utilise human-wastes. Before putting Wardha-type laterines in use, a 7.5 cm layer of refuse (grass, leaves, straw, weeds, trash, etc.) is spread at the bottom of the trench. A stock of refuse or dry earth is left to cover the faeces. After the material rises to a height below 15 cm of the top, the trench is sealed for decomposition and another trench put in use. Manure of good quality is ready in four to six months.

Preparation of Synthetic Compost

The basic principles underlying the microbiological process of conversion of organic wastes into manure have been extended for preparation of synthetic composts, using inorganic fertilizers as sources of nitrogen for decomposition of carbonaceous materials. It has been found that organic nitrogen in the form of dung, required by micro-organisms, can be completely substituted by inorganic nitrogen compounds like ammonium sulphate and urea, which are utilised equally effectively for decomposition and well fermented manure, resembling farmyard manure, can thus be obtained. This ;pens up a vast field for utilisation of large quantities of various organic waste materials where supplies of dung are either short of requirements, or not available at all, as on mechanized farms. The b ask principles of C/N ratio in manure preparation can be applied to add nitrogenous fertilizers in sufficient quantity to reduce the ratio to about 30-40:1 and then allow the material to be decomposed and disintegrated until the C/N ratio is reduced to 20:1. The adco process of preparation of synthetic compost worked out by Hutchinson and Richards is based on this principle.

This process of production of compost from organic wastes of wide C/N ratio is worked with fertilizers like ammonium sulphate or urea. They provide complete or partial nitrogen for microbial decomposition in place of natural materials like dung or night-soil which may not be available at all or may be in short supply. Nitrogen is added to make up a total of 1.2 per cent on dry weight basis of the material. Thus straw containing 0.4 per cent nitrogen requires an additional 0.8 per cent nitrogen, *i.e.* about 4 kg of ammonium sulphate per 100 kg of straw. For neutralizing acidity, which may develop in this system, the process includes addition of lime at the rate of 5 kg to the above material. The material to be composted is spread out in layer in a heap or pit and sufficiently moistened. This is sprinkled with the fertilizer solution and then with lime. Superphosphate may be added to fortify the phosphatic contents of the manure. The treatment is continued layer-wise until the heap or pit is filled to size and allowed to ferment. The manure becomes ready for application in about four to six months and resembles farmyard manure in its action on soil and plant growth.

Mechanical Compost Plants

Although the major quantities of compost (310-350 million tonnes) at present used in agriculture are prepared in villages either by traditional or improved methods (Indore or Bangalore methods), these methods cannot be suitably used for processing of large quantities of organic refuse of bigger cities.

The mechanised/semi-mechanised plants will serve two pertinent objectives- sanitary disposal of city refuse and production of organic manure. It is estimated that by 1979 nearly 50 million tonnes of municipal wastes will be generated annually in India. Moreover, Indian city wastes contain about 50 to 80 per cent of compostable organic materials which can be processed conveniently in compost plants. During Sixth Five Year Plan, Ministry of Agriculture and Irrigation timely launched a programme to set up 35 composting plants in selected cities with a population of 3 lakhs and above. Five compost plants have already been set up one each at Ahmedabad, Baroda, Bangalore, Kolkata and Jaipur and the remaining 30 plants are expected to be set up during Sixth Five Year Plan (1978-83). The intake capacity of these plants already set up is about 125 to 200 tonnes garbage per day and can produce 60 to 70 tonnes of finished compost per day, if the plant runs in shifts.

The costs of production is rather high due to mechanisation of the processes and at some places the plants were running at under-utilised capacity due to non-availability of required quantities of garbage. The selling price of finished compost has been kept between Rs 50 to 80 per tonne depending upon the composting system adopted and plant capacity. The cost may be reduced by installation of plants with less mechanisation.

The compost prepared from city wastes is bulky and low in plant nutrients. It contains 0.5 to 0.6 per cent N, 0.6 per cent P_2O_5 and 0.5 per cent K_2O with a C/N ratio of 20:1. There is a need to enrich this manure in respect of nitrogen and phosphorus and reduce its bulkiness involving low-cost technology. The work carried by Gaur showed that by application of 10 kg low-grade rock phosphate containing 20 per cent P_2O_5 per tonne of organic wastes (crop wastes and leaf fall, etc.) and inoculation with microbial fertilizers improved the nitrogen, phosphorus and humus content appreciably. This can be adopted in improving the compost prepared by mechanical compost plants.

Use of Composted Versus Uncomposted Materials

Direct application of carbonaceous materials of wide C/N ratio to soils adversely affects plant growth. This is owing to the utilisation of all soil nitrogen by micro-organisms, engaged in the process of decomposition. Mixed organic wastes containing adequate nitrogenous materials like dung and night-soil can, however, be directly applied to land and allowed to decompose *in situ* in soil without causing deficiency of nitrogen and the consequent depressing effect on plant growth. If the C/N ratio of the materials is restored to 30:1 or less, the decomposition in soil is faster owing to dilution and large variety of microbial population of the soil. Positive advantages that accrue from the decomposition carried out in the soil are: (a) no loss of nitrogen in the form of ammonia takes plate in the presence of the large quantity of soil; (b) carbon dioxide evolved in large quantities during decomposition assists in enhancing the available plant nutrients of the soil; and (c) production of compounds possessing soil aggregating properties during decomposition by micro-organisms, imparts good structure to the soil. The succeeding crop thus derives greater benefit from this manner of application than from the same raw materials after composting. It is, however, obvious that there are some limitations to the use of raw waste materials in practice. Materials like long-straw and street-refuse may be difficult to incorporate without prior disintegration. Offensive materials like night-soil cannot be handled in raw condition. Composting destroys pathogenic organisms and weed seeds that may be present in such materials. Under intensive cropping, the gap between one crop and the other may be too short for direct application of organic materials to soil and they may have to be composted before application. It has, however, often established that whenever possible, direct incorporation of materials like straw, leaves, etc. sufficiently in advance of sowing, with supplementary application of fertilizer nitrogen in adequate quantity, results in marked improvement in soil fertility with significant effect on crop yield.

Night-Soil Poudrettes

The disposal of night-soil is often a serious problem in towns where facilities for its composting are not available or refuse in adequate quantity is not collected. Trenching the material as such takes a long time for it to dry and causes nuisance of foul smell and profuse fly-breeding. Dehydration of night-soil by mixing it with any absorbing material like soil, ash, dry broken leaves, paddy-husk, etc. is a highly effective method of disposal of this offensive waste which yields poudrette manures of high manurial value. One of the most effective materials for this purpose is sawdust, which has high dehydrating capacity as well as the property of absorbing foul smell. The material can be used whenever available to treat night-soil and obtain an organic manure of high quality. Owing to its low C/N ratio, utilisation of night-soil as such without fermentation yields higher efficiency of response than when it is composted. The question of survival of pathogenic organisms in preparation and use of poudrette manures of this type is, however, a matter warranting consideration and further study.

Super-Digested Compost

Recent trends in the preparation of farmyard manure and composts include the use of superphosphate which is sprinkled over the raw materials as they are being filled in pits, at the rate of about five per cent. It has been found that the treatment makes organic manures of this type more balanced and markedly reduces losses of nitrogen that, take place during decomposition. The availability of phosphorus, thus intimately mixed with organic matter, is higher than when this fertilizer is used as such and this is of particular advantage to acidic, calcareous and heavy soils in which fixation of phosphorus greatly reduces the efficiency of this plant nutrient. Results of work by Walunfkarhar and Acharya illustrated the effect of super-digested compost on crop growth are given in the Table 2.9. Yield of *mania* was significantly better with application of super-digested compost than with separate application of the two.

Table 2.9: Average Yield of *Mama* in g per Pot

S.No.	Treatment	Grain	Straw	Total
1	No manure	7.4	23.8	31.2
2	Compost (2 ½ tonnes dry matter/hectare)	10.3	33.0	43.3
3	Superphosphate (P_2O_5 equal to No. 2)	8.9	30.8	39.7
4	Super-digested compost	12.3	41.6	54.1
5	Compost plus superphosphate (as in No. 4)	10.3	32.6	42.9
	S.E.	0.35	1.73	3.18

Subbiah by the use of tracer technique, showed that the movement of applied phosphorus in soil generally increased, when organic matter in the form of compost was also applied with the fertilizer and this was further enhanced when applied as super-digested compost. The increase in movement due to super-digested compost over mere combination of superphosphate and compost was higher in laterite, red and calcareous soils.

Control of Losses of Nitrogen During Manure Preparation

It has been mentioned that the process of decomposition in the preparation of organic manures results in significant losses of nitrogen in practice, which reduces the quality of final manure that is obtained. The losses ranging upto 20 per cent take place although they are heavy in heaps and are considerably reduced in pits. It is, therefore, always desirable to prepare manure in pits to obtain a better quality and quantity of manure.

Loss of nitrogen during fermentation takes place mainly in the form of ammonia gas and methods have been worked out to retain this in the manure by the use of acidic materials to fix nitrogen chemically or by absorbants which physically hold nitrogen. Thus, addition of superphosphate at the rate of about five per cent or a heap mineral acid like hydrochloric acid at one per cent have been found to conserve nitrogen in the manure. Materials like charcoal-dust, sawdust, gypsum and even soil added at the rate of five to ten per cent reduce nitrogen losses considerably. The inclusion of this simple treatment in the process of manure preparation would result in obtaining a superior quality of manure in which a substantial amount of nitrogen is held in forms that are highly available to plants.

Enriched Organic-based Manures

Fermented organic manures are low grade materials and, under Indian conditions, have to be applied in large bulk to obtain significant effects on crop growth and yield. The total nitrogen content in them does not exceed 1.5 per cent at best and with the small quantity of dung used by farmers, it is of the order of 0.5 per cent. The percentage availability of nitrogen for plants in such manures ranges from 5 to 20 per cent with the result that low responses are obtained, when applications are made which are inadequate to make good the primary deficiency of nitrogen in our soils.

At the Indian Agricultural Research Institute, a new approach to this problem has been worked out which enables such low-quality manures and other organic materials to be effectively utilized. It has been found that the rate and extent of liberation of nitrogen from applied organic manures is adequate for normal plant growth, when their C/N ratio is less than 10 :1 or the nitrogen content is more than 2.5 per cent. In practice, the C/N ratio of organic materials can be reduced by adding the requisite amount of nitrogen in the form of any nitrogenous fertilizer to it and thereby increasing the value of the denominator instead of decreasing carbon in the numerator by fermentation. Thus, a low-quality manure containing 0.5 per cent nitrogen and a C/N ratio of 20:1 can be treated with ammonium sulphate or urea solution to add 2 per cent more nitrogen and raise the total nitrogen content to 2.5 per cent. This would immediately reduce the C/N ratio to 8 : 1 and the enriched manure can then be utilised effectively. This method can utilise large quantities of other organic waste materials of wide C/N ratio like dry dung, paddy and groundnut-husks, sawdust, tobacco-waste, wheat-chaff and any material in condition to be easily incorporated in soil. This method enables the process of composting to be eliminated, thus, saving time and labour.

Enriched organic manures of the above type are processed and get ready in a short time and possess many advantages. Losses resulting from their composting are avoided and the total quantities of organic matter and nitrogen taken initially are conserved and utilised. The decomposition taking place in the soil in the presence of adequate nitrogen incorporated will cause no adverse effects and sufficient nitrogen, in excess of microbial requirements, is liberated for optimum growth of crops.

Variations in the preparation of these enriched manures include incorporation of more nitrogen to obtain concentrated manures containing 5 to 7 per cent nitrogen and supplementing this with superphosphate and potassium sulphate to get mixed organic-mineral fertilizers of any desired grade to suit specific requirements of soil or crops.

Preparation of Enriched Manures

Enriched organic-based manures are prepared by taking requisite quantities of nitrogenous and other fertilizers and dissolving them in sufficient water. The solution is then treated with 100 ml of commercial hydrochloric acid per quintal of the final manure for controlling decomposition of the added nitrogen and its retention in the manure. This is then absorbed in the calculated quantity of the organic material, mixed thoroughly and dried in the sun. The dry enriched manure can then be bagged and stored for use.

Microbial Method for Hastening Compost Preparation

Compost is bulky in nature and contains on an average 0.5 to 1.0 per cent N, 0.6 pet cent P_2O_s and 0.5 per cent K_20. There is a need to reduce its bulkiness and period of composting to make it more acceptable to the farmer. Moreover, the quality of manure should be improved to supply more nitrogen and phosphorus and humus content per unit weight. Researches conducted by Gaur in the Division of Microbiology, IARI have shown that it is feasible to hasten the period of composting by proper inoculation with cellulolytic micro-organisms as well as improve nitrogen and phosphorus contents of organic fertilizers with low-cost technology. Gaur reported that mesophilic cellulolytic fungi, when used as inoculant in composting of plant residues (straw, leaf fall) can hasten the composting process and improve the quality of compost. The carbon content of compost mass was lowered by 25 per cent of the initial raw material (48 per cent) organic carbon during the first month of composting in pits. The loss was more due to fungal inoculation, *Penicillium* spp. showing the maximum decomposition of 46 per cent followed by *Aspergillus* spp. with 37.5 per cent. As a result, C/N ratio within one month was decreased ranging from 19.6 to 23.6 due to inoculation with cellulolytic fungi, whereas in uninoculated control, it was 31.4 (Table 2.10). Subsequently also the decreases in C/N ratio were greater due to inoculation.

Table 2.10: Effect of Cellulolytic Culture in Hastening the Process of Composting

Treatments	C:N ratio of compost sampling interval in months		
	I	II	III
Control	31.4	22.4	20.0
Aspergillus sp. (IF_1)	23.6	-	16.8
Trichoderma viride (IF_2)	23.1	19.5	18.9
Aspergillus sp. (RF_1)	20.6	16.9	14.3
Penicillium sp. (RF_2)	19.6	16.8	14.8

Initial C/N ratio 45.7; Weight of raw material 32 kg/pit

The data (Table 2.11) on recovery percentage of compost on oven dry basis indicated that bulkiness : f organic manures decreased by 2 to 8 per cent due to inoculation. The efficient culture reduced bulkiness by 8 per cent. The total nitrogen content showed increase on the other hand, the increase being more in inoculated series. Maximum of 17 per cent increase was recorded with *Aspergillus* spp. (RF_1) followed by *Aspergillus* spp. (IF_1) and *Penicillium* spp. (RF_2) with 15.6 and 14.2 respectively as compared to uninoculated control.

Table 2.11: Final Yield of Compost and Nitrogen Status in Compost Material

Treatment	Final yield of compost (kg)	% Recovery	Total N %	Total N in Finished compost material (g)	Total N in Per tonne of finished compost (kg)	% Increase over control
Raw material	-	-	1.05	(336)	(10.5)	-
Control	19.6	61.25	1.41	276	14.1	34.29
Aspergillus sp. (IF_1)	18.9	59.06	1.53	308	16.3	55.24 (15.6)
Trichoderma viride (IF_2)	18.1	56.56	1.33	241	13.3	26.67 (-5.6)
Aspergillus sp. (RF_1)	17.0	53.12	1.65	281	16.5	57.14 (17.0)
Penicillium sp. (RF_2)	16.9	52.81	1.61	272	16.1	53.33 (14.2)

Weight of organic material composted 32 kg/pit
Figures in parentheses show increase of decrease over uninoculated control.

The studies indicated that due to inoculation of compost heaps by cellulolytic cultures the period of composting of dry and wide C/N ratio materials can be reduced by 4 to 6 weeks. These findings have been confirmed by the studies carried out at other centres.

Lost Cost Technology for Enrichment of Compost

Recent experiments conducted at IARI have shown that inoculation with *Azotobacter chroococcum* and phosphate solubilising micro-organisms *(Aspergilus awamori* or *Bacillus polymyxa)* improved the manurial value of compost. Inoculation of compost with nitrogen fixer and P solubiliser was done, when the temperature had stabilised around 30-35°C after one month of composting. Low-grade rock phosphate (10 kg/tonne of raw material) was applied at the initial steps of filling up of compost pits.

The results indicated that 31.5 per cent carbon was decomposed in control during the first month, whereas the loss in different treatments ranged between 33.3 to 41.5 per cent. This was further increased during second month showing maximum loss of 52.4 per cent of organic carbon in rock phosphate, *Azotobacter* and P solubilizer inoculated series (Table 2.12). In this treatment C/N ratio was reduced to 14.1 and 12.0 during 1st month and 2nd month after inoculation respectively.

Table 2.12: Changes in Carbon (%), Total Nitrogen (%), C/N Ratio During Enrichment of Compost by *Azotobacter* and P Solubilizing Cultures (Sampling interval in months after inoculation)

Treatment	1st Month				2nd Month			
	Carbon	Nitrogen	C/N Ratio	Humus %	Carbon	Nitrogen	C/N Ratio	Humus %
Control	31.4	1.36	23.0	7.8	28.5	1.38	20.6	4.85
Control + Rock Phosphate @ 10 kg P_2O_5/tonne	30.7	1.61	19.0	9.15	27.5	1.77	15.5	6.90 i
Azotobacter + P Solubilizer	30.5	1.51	20.1	9.30	26.6	1.73	15.4	6.00
Rock Phosphate + Azotobacter + P Solubilizer	26.9	1.89	14.1	9.0	21.9	1.82	12.0	6.95

Initial sample: Organic carbon 46.0%, Total nitrogen 0.86%, C/N ratio 53.3

There was appreciable gain in nitrogen content per tonne of finished compost in control as well as inoculated series ranging from 60.5 to 111.6 per cent over the initial raw material (Table 2.13). Addition of powdered rock phosphate alone resulted in an increase of 28.3 per cent over control. Inoculation with *A. chroococcum* and *A. awamori* amended with rock phosphate showed an increase of 32.0 per cent in nitrogen content over the control resulting in gain of 22 kg N/5 tonnes of finished compost. It was also noted that humus content increased in all the treatments including control (Table 2.11). The best effect in humus content 9.3 per cent was observed in case of *A. chroococcum* + *A. awamori* after 1 month of inoculation.

In another study the number of *Azotobacter chroococcum* in compost was increased due to its inoculation and use of rock phosphate. The mean numbers of nitrogen fixers of nine weekly intervals per gram of dry compost in different treatments were as follows: Control 1,380, rock phosphate 3,590; *Azotobacter* 6,070; rock phosphate + *Azotobacter* 21,010. This also resulted in an increase in the nitrogen content of conlpost by 25 to 44 per cer.t over control. These findings clearly show that quality of compost can be improved by inoculation with *Azolobacter* and P solubilisers suitably amended with rock phosphate.

Table 2.13: Final Yield of Compost and Nitrogen Status in Compost Material

Treatment	Weight of Raw Material (kg/pit)	Final Yield of Compost (kg)	% Recovery	Total N%	Total N in Finished Compost Material (g)	Per Tonne of Finished Compost (kg)	% Increase Over Control
Raw material	32	-	-	0.86	275	8.6	-
Control	32	11.85	37.03	1.38	164	13.8	60.46
Rock phosphate @ 10 kg P_2O_5/ tonne	32	12.30	38.44	1.73	213	17.3	101.16
Azotobacter + P solubilizer	32	11.45	35.78	1.77	203	17.7	105.81
Rock phosphate + Azotobacter + P solubilizer	32	13.00	40.63	1.82	236	18.2	11.03

Fly-Breeding in Manure Preparation and its Control

One of the disturbing aspects of manure preparation from organic-wastes particularly town-wastes, is the profuse fly breeding that takes place in this medium. In over-ground heaps, owing to the heat that develops inside, fly maggots migrate and concentrate on the surface few cm all over. If the material is given turnings, as in the Indore method, the maggots get substantially destroyed, otherwise flies develop profusely. In pits, the top layer of refuse or soil becomes the seat of migration of all maggots where they finally develop into flies.

The emergence of flies from such manure pits has been observed from the 6th day after filling and goes on till the 12th day or so, after which it ceases. Based on these observations, a simple and highly effective method of destroying flies, while they are resting in the maggot or pupae stage in the surface layer, has been worked out. The entire surface area of the pit is covered on the 6th day after filling with rags, refuse, straw, cotton or other stalks, sawdust, paddy or groundnut-husks or any such combustible waste and set on fire, which may continue to smoulder for about an hour. The maggots and pupae lodged on the few centimetres of upper surface are completely destroyed by the heat. The operation may be repeated on the 12th day to obtain absolute control of fly breeding. The treatment leaves a heat layer of ash on the top which is no more attractive to the flies.

Oil-Cakes

Oil-cake is a valuable by-product of the oilseed extraction industry and contains significant quantities of proteins, of value as animal feed, if edible and as nitrogenous manures otherwise. There is always a residue of oil present in them depending on the process of extraction as shown below:

	Oil per cent
Country ghani extracts	10-12
Hydraulic press	8-10
Stream expeller	5-8
Solvent extract	1-2

The presence of oil slows down decomposition and the process of mineralization of their nitrogen in soil. Since this also gets decomposed in due course, the overall formation of mineral nitrogen is not materially affected in quantity. Their C/N ratio is narrow varying from 3 to 10 and about 60 to 75 per cent of the nitrogen is easily mineralizable and available for plants.

Effective Utilization of Mahua-Cake

Mahua-cake is the poorest of oil-cakes with a nitrogen content of 2.5 per cent and has been found to be much less effective than others even when applied in larger quantities. It has been found that nitrogen of this manure mineralised to a small extent in the soil owing to the presence of toxic saponins. The efficiency of the manure can be increased by moitening it and allowing it to ferment for some time to decompose the toxic substances. An effective treatment worked out at the LARI involves treatment of the material with ammonia solution which not only increases the nitrogen content of the manure but also enables the original nitrogen to be effectively mineralized and utilized by plants. The treatment could be extended to enrich other oil-cakes and utilize them more effectively.

Utilization of Slaughter-House Wastes

Blood

Blood is the most valuable of slaughter-house wastes which yields a rich proteinaceous meal on dehydration. It is useful as a feed or as a nitrogenous manure. The existing crude methods of its collection and processing in India result in wastage of a large part.

In slaughter-houses, blood is obtained partly in the form of spongy clots and partly as blood serum in liquid condition. These are dried by heating in large open pans over a slow fire, which is a time-consuming and highly offensive process. The Indian Agricultural Research Institute has worked out a simple method for quick treatment of blood to obtain blood-meal which can be used as feed or as a manure.

Blood-meal from serum is prepared by adding one per cent of lime and then heating it till the temperature rises to 80°C. At this stage, blood begins to coagulate by its reaction with lime, and further slow heating is continued with constant stirring until the flocculated blood separates out and the colour changes from red to ash. The remaining water is decanted off and the coagulated blood is passed through a quarter-inch mesh expanded metal sieve and dried in the sun. There is no loss of nitrogen in the process and the manure contains 10 to 12 per cent nitrogen.

Blood clots are processed separately by cooking them for a few minutes boiling in 0.5 per cent hydrochloric acid solution. The clots harden and rise upto the surface. They are then taken out, rubbed over an expanded metal seive and dried. The manure contains 14 to 15 per cent nitrogen.

Blood serum can also be absorbed as such in materials like dry dung powder, sawdust, dry leaf powder and such other materials and then dried in the sun. Such products may contain 4 to 5 per cent nitrogen. The use of these absorbing materials on the floor of slaughter-houses would enable blood to be directly absorbed and offers one of the simplest methods of effective collection and utilisation of this waste and keeping slaughter-houses clean.

Waste-meat

This waste in slaughter-house consisting of intestines, offal and other flesh cuttings is hardly utilized in our country. Their disposal and use in compost pits has been suggested where they can serve as a source of nitrogen required for the process of decomposition of other wastes. The proposition of utilization of meat wastes as such and obtaining concentrated meat-meal is more useful to consider. The method of cooking these in steam digesters under pressure can be adopted for the purpose but this requires expensive equipment. A simple method worked out at the IARI for treatment of such materials consists of cutting them into pieces and cooking in 5 per cent hydrochloric acid solution until they become soft. The material is then washed and rubbed over an expanded metal sieve and dried in the sun. The product may contain 8 to 10 per cent nitrogen and can be used as poultry feed or manure. The cost of processing works out to Rs 2.50 per tonne.

The method can also be adopted to utilize the entire flesh of carcasses of all animals that die in towns and villages and can be organized on a countrywide scale for the exploitation of a valuable source of manure.

Bones

Bones from slaughter-houses, carcasses of all animals and from meat processing industry constitute a valuable waste which could be exploited more fully in the country. The existing methods of disintegration of bones involve crushing them in hammer mills by which raw bone-meal is obtained as a by-product or by digesting them in steam digesters, under pressure which yields steamed bone-meal in fine powdery condition. The methods require expensive equipment and are not always suited for adoption and treating small collection in villages and remote places.

Chemically Processed Bone-Meal

The IARI has evolved a purely chemical process of disintegrating bones and obtaining a sterilized bone-meal in fine condition which can be used as a feed or as manure. The process can be worked both as a cottage industry to treat small collections, suiting conditions in this country or on a large scale, if required.

Fresh or dry bones are taken in a suitably-sized container and soaked in sufficient water to cover them. Caustic soda is added to this to make a 5 per cent solution in the quantity of water taken. The bones are stirred once or twice daily until they disintegrate completely to a paste in 14 days. The supernatent is then decanted and utilized again and the disintegrated bones are washed twice with water. The material is finally taken up with water and treated with 10 per cent hydrochloric acid solution until it is neutral to litmus paper. The supernatent is then decanted and the bone paste filtered through muslin cloth. This is then dried in the sun and it yields a sterilized bone-meal in fine condition. The product contains about one per cent nitrogen and 27 per cent phosphoric acid. The cost of production, excluding the value of bones works out to be Rs 60 per tonne. The process is patented.

Fish-Meal

Fish manure processed by drying non-edible fish or wastes from fish industry is a well-recognized, balanced organic manure containing both nitrogen and phosphorus in significant quantities. Fish-meal is also produced in India by cooking in steam digesters and then drying. The manure, however, has an extremely offensive smell which stands in the way of its wider use. At the IARI, it has been found that the objectionable smell can be completely smothered by the use of two litres of pipe-oil emulsion per tonne. The long coast-line of the country offers a great scope for fuller exploitation of this highly valuable manure for increasing agricultural production.

Leather-Wastes

Tanned and untanned leather wastes are obtained from tannery and leather processing industries where their disposal becomes a serious problem. Leather-wastes can be processed by heating in steam digesters and then grinding to powder. The product obtained contains 7 to 8 per cent nitrogen and has been used as a nitrogenous manure. The IARI has a patented process for preparing leather-meal in which the material is soaked in water and then cooked in 5 per cent hydrochloric acid. The leather disintegrates into a plastic mass which is washed in water and dried and ground to powder. The acid treatment also helps in breaking down complex nitrogenous compounds and yields a manure containing 7 to 8 per cent nitrogen of which 70 to 75 per cent is mineralizable. The method can be adopted on a cottage industry scale, or large-scale according to the needs.

Hair and Wool-Wastes

Large quantities of animal-hair wastes become available in tanneries, which have found no use so far. Human hair can be collected in a large quantity, from hair-cutting saloons. These have also not been exploited so far. Large quantities of wool-waste also become available from wool refining, cloth and carpet industries which are not properly utilized. These materials contain about 15 per cent nitrogen in the form of complex keratin and do not decompose in soil, when used as such. No estimate is available of the quantities of these wastes that can be collected but it is considered that their utilization in agriculture as manure after suitable processing would provide a substantial additional source of a rich organic manure. A patented process worked out at the IARI involves treatment of these materials with a solution containing 8 per cent caustic soda on the weight of the material. This brings about their disintegration to a paste which is then treated with 10 per cent hydrochloric acid to neutralize the alkali and re-precipitate the proteins. The liquid is decanted and the spongy solid is rubbed over an expanded metal sieve and dried. The product contains 12 to 15 per cent nitrogen and has been found to be a good nitrogenous manure.

3

Development of Biofertilizers

At present India produces about 117 million metric tonnes of food grains for its growing population of more than 800 million people. During the recent plan the target to produce is additional 30 million food grains and 5 million tonnes of pulses and oilseeds. In order to meet the food requirement of ever increasing population, the nitrogen fertilizer requirement for crop production by 2000 AD is estimated to be about 11.4 x 10^6 tonnes as against 3.9 x 10^6 tonnes. At present the waste gap can not be filled up merely through the production of synthetic nitrogenous fertilizers due to scarcity of high cost of raw materials, such as fossil fuels. Biological nitrogen fixation is the key to sustain agricultural productivity application of biofertilizers in the field is the viable alternative. Biofertilizer as a living fertilizer composed of microbial inoculants or groups of micro-organisms which are able to fix atmospheric nitrogen, the micro-organisms are known as biological nitrogen fixers. They are grouped into free living bacteria *(Azotobacter* and *Azospirillium)* and the blue green algae and symbionts, such as *Rhizobium,* and *Frankia* and *Azolla.*

In fertilizer manufacturing factories, nitrogen is fixed industrially, by means of the Haber-Bosch process requiring H_2 gas, very high temperature and anormous energy. Industrially fixed nitrogen has been used precipituously. It was produced 4.0 lakh metric tonnes in 1905. This was increased to 3.5 million tonnes of nitrogen fertilizer was produced industrially. Meanwhile the contribution of biologically fixed nitrogen has not changed. The estimate of chemically fixed nitrogen for 1976 is 42 million metric tonnes. It is projected that by 2000, if we continue to rely entirely on increased use of chemical fertilizers to achieve the food production level more demand for fertilizer nitrogen to results and around 100-200 million metric tonnes would be needed to produce at an average cost of rupees.

Materials of Biological Origin

Indian soils are usually poor in organic matter as well as in nitrogen. The materials of biological origin which are commonly used to maintain and improve soil fertility may be grouped into two main categories:

1. Biofertilizers and
2. Green manures

The term biofertilizers or which can be more appropriately called 'microbial inoculants' can be generally defined as a preparation containing live or latent cells of efficient strains of nitrogen fixing, phosphate solubilising or cellulytic micro-organisms used for application of seed, soil or composting areas with the objective of increasing the numbers of such micro-organisms and accelerate certain microbial process to augument the extent of the availability of nutrients in a form which can be easily assimilated by plant. In large sense, the term may be used to include all organic resources (manure) for plant growth which are rendered in an available form for plant absorption through micro-organisms or plant associations or interactions.

Classification

Biofertilizers can be broadly classified into three categories (Table 3.1) :

Table 3.1: Classification

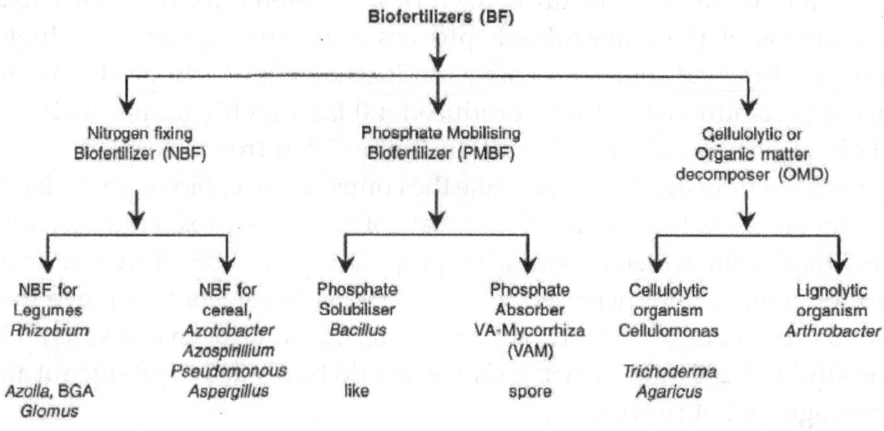

Potential of Biofertilizers in Crop Production in Indian Agriculture

Soil harbours a range of micro-organisms which bring about a number of biochemical reactions in soil. The soil organisms are classified into two broad groups *i.e.*, soil flora and soil fauna; these are again subdivided, depending upon their size, such as micro and macro flora. Soil microflora includes bacteria, fungi, actinomycetes, algae, etc. Of these groups, bacteria are the most abundant followed by actinomycetes, and fungi; algae are found under specific situations. The biomass and population of these micro-organisms in soil are found under specific situations. The biomass and population of these micro-organisms in soil have been given in the Table 3.2.

Table 3.2: Important Soil

Micro-organisms	Average Number (in lakhs) per g of Soil	Average Biomass (in kg per ha)
Bacteria	1000	500
Fungi	10	1000
Actinomycetes	100	750
Algae	0.01	150

The presence of these micro-organisms make the soil a living and active system. These micro-organisms play a significant role in the life cycle of plants and animals through a number of processes, such as decomposition, solubilization, nutrient fixation and supply of plants (Table 3.3).

Biofertilizer means using living organisms as fertilizer, either to fix atmospheric nitrogen or to solubilize mineral nutrients like phosphorus. The microbial inoculants have attained special significance in modem agriculture. Keeping in view the increasing cost of chemical fertilizers & poor purchasing capacity of Indian farmers several micro-organisms which can be used in agriculture are being listed in Table 3.4.

Chemically Fixed Nitrogen Versus Biologically Fixed Nitrogen

With the fast depletion of fossil fuel and exorbitant cost of chemical fertilizers, efforts should be made for augmenting biological nitrogen fixation by micro-organisms. On an average, one acre of grain legumes (like soybean, bean or pea) provide sufficient protein for 1000-2000 days for one person, whereas an average acre of plant materials converted to animal protein like beef and poultry provides it only for 75-250 days.

India being one of the less affluent nations, where the per capita income is low and consequently, the intake of vegetable protein is more in comparison to animal protein. Accurate estimates of annual turnover of nitrogen in the biosphere vary from 100 to 200 million tonnes. The amount of nitrogen fixed by various legumes irrespective of soil and other conditions has been given in the Table 3.5.

Table 3.3: Important Biofertilizers Agents

Group	Scientific Name	Activity
Bacteria	*Bradyhizobium* sp.	Symbiotic nitrogen fixation
	Rhizobium sp.	Symbiotic nitrogen fixation
	Azotobacter sp.	Asymbiotic nitrogen fixation
	Azospirillium sp.	Asymbiotic fixation
	Pseudomonas striata	Phosphate solubilization
	Bacillus subtilis	Phosphate solubilization
	Xanthomonas sp.	Phosphate solubilization
	Nitrosomonas sp.	Phosphate solubilization
Fungi	*Aspergillus niger*	Phosphate solubilization
	A. awamori	Phosphate solubilization
	Fusaruim sp.	Phosphate solubilization
	Penicillium sp.	Phosphate solubilization
Vesicular arbuscular mycorrhizae (VAM)		Phosphate solubilization
Algae	*Anabaena, Nostoc, Calothrix*	Biological nitrogen fixation
Fern	*Azolla* sp.	Nitrogen fixation
	A. caroliniana	Nitrogen fixation
	A. nilotica	Nitrogen fixation

Table 3.4: Important Micro-organisms Constituting Biofertilizers

Microorganism	Nutrient Fixed (kg ha//year)
Actinorrhizae (*Frankia* sp.)	150 kg nitrogen ha
Algae	25 kg N_2/ha
Azolla	900 kg N_2/ha
Azospirillum	10-20 kg N_2/ha
Rhizobium	50 to 300 kg N_2/ha
Azotobacter	0.026 to 20 kg N_2/ha
Mycorrhizae	Solubilize food phosphorus (60 per cent)
Phosphate solubilizing bacteria and fungi	Solubilize about 50-60 per cent of the fixed phosphorus in the soil

Table 3.6: Estimates of Nitrogen Fixed by Some Legumes

Crop	Nutrient Fixed (kg per hectare)
Alfalfa	100-300
Berseem	120-150
Chickpea	26-63
Clusterbean	37-196
Clover	100-150
Cowpéa	53-85
Blackgram	38-50
Greengram	50-55
Groundnut	112-152
Lentil	35-100
Peas	59-80
Pigeonpea	60-200
Soybean	49-130
Sesbania	69-90

Above every hectare of land, there are 78,000 tonnes of inert nitrogen gas (N_2). Nitrogen is the most limiting nutrient for increasing crop productivity. Only a few procaryotic organisms are able to fix nitrogen directly through a biological process. Annual biological nitrogen fixation (BNF) is estimated to be around 175 million tonnes of which close to 79 per cent is account for by terrestrial fixation.

Synergistic Interaction Between Biofertilizing Agents

Biofertilizing agents, such as *Rhizobia,* mycorrhizae, *Azotobacter, Azospirillum, etc.* are member of soil-biota or soil biosphere; they live together since the introduction of life in the universe. Hence, they have developed a sort of natural relationship with each other and they maintain it unless disturbed by human beings by growing crops, application of fertilizers or pesticides. Crop growth in the soil releases several organic substances which may selectively eliminate certain micro-organisms or reduce their activities or even augment the activities of the soil biota. Since various biofertilizing agents are selective, such as *Rhizobia* fix N_2 from atmosphere in the root zone of the legumes or phosphate solubilizing bacteria (PSB) or fungi (PSF) solubilize phosphate in the soil; this results in the simultaneous application of *Rhizobia* and PSF or mycorrhizae. The question is always raised whether the simultaneous application may not jeopardize the activity of the individuals. Experiments were conducted on the simultaneous application or blending application of biofertilizing agents and it is generally concluded that the application has, by and large synergistic action (Table 3.6).

Table 3.6: Interaction of Various Biofertilizing Agents

Crop	Inoculation of biofertilizing agent	Positive (+) or Negative (-)
Wheat	*Azotobacter* (AZOTB + *Rhizobium* (Rh)	+
Peanuts	*Azospirillum* (AZOS) + *Glomus* sp. + *Rhizobium*	+
Onion	*Beijerinckia mobilis* + *Aspergillus niger* + *Glomus fasciculatum*	+
Tomato	*Azotobacter vinelandi* + *G. fasciculatum* + *Rhizobium* sp.	+
Chickpea	*G. caledonicum* + *Rhizobium*	+
Soybean	*Glomus* + *Gigaspora* + *Rhizobium* sp.	+
Grases	*Azotobacter brasilense* + *Glomus* sp.	+

Biofertilizers are apparently environment and farmers friendly renewable source of non-bulky, low cost organic agro-input. While *Rhizobium, Blue Green Algae* (BOA) and *Azolla* are crop specific, bio-inoculants like *Azotobacter, Azospirillum, Phosphorus Solubilizing Bacteria* (PSB), *Vesicular Arbuscular Mycorrhiza* (VAM)etc. could be regarded as broad spectrum biofertilizers. Their use has so far not jeceived desired attention. This is mainly because of inadequate awareness both among the extention workers and farmers regarding their utility, their short shelf life, lack of ready availability in time and in desirable quality, and inconsistency in results with their use. However, proper education of the extension workers, dealers and farmers about their significance and economic feasibility of their application both by seed treatment or to soil, could go a long way in prompting their adoption. These could effectively supplement the nutrient requirement of crops through chemical fertilizers, to meet the soaring demand for food, fibre and fuel.

A biofertilizer is an organic product containing a specific microorganism (microbial inoculant) in concentrated form (10^7 to 10^9 per gram) which is derived either from the nodules of plant roots or from the soil of root zone (rhizosphere). They possess unique ability to fix atmospheric nitrogen either by living symbiotically with the roots of leguminous plants or non-symbiotically (free living) or to transform native soil nutrients, such as phosphorus, zinc, copper, iron, sulphur, etc. from the non-usable (fixed) to usable form through biological processes. Biofertilizers are apparently environmental friendly, low cost, non-bulky agricultural input which could play a significant role in plant nutrition as a supplementary and complimentary factor to mineral nutrition.

Biofertilizing Agents and Plant Disease Control

Biofertilizing agents control the plant pathogenic fungi directly as well as indirectly. Directly they parasitize the pathogens: application of rhizobium culture on the legume seeds control seed borne fungi, such as *Colletotrichum, Ascochytii, Helminthosporium* etc. The rhizobia produce a toxic substance when they multiply on the seed and rhizosphere. Phosphate solubilizing fungi such as *Aspergillus niger* and other *Penicilla* produce antibiotic substances and thus, kill the pathogenic fungi. Indirect killing of the plant pathogens is achieved by producing healthy seedlings and phytoalexins. Application of mycorrhizae produce better root systems which overcome the attack of root rotting and soil borne pathogens. Numerous reports are available that applications of biofertilizers in the soil stimulate and augment the activity of saprophytic micro-organisms.

Biofertilizers and Insect Control: Legume inoculation indirectly protect plants from the attack of insects.

Brief Account of Beneficial Microorganisms

Rhizobium

Symbiotic N_2 fixation by *Rhizobium* in legumes contribute substantially to total biological nitrogen fixation. Rhizobium inoculation is well known agronomic practice to ensure an adequate nitrogen of legumes in lieu of N-fertilizer. Different species of rhizobium are classified into two groups *viz.,* (i) slow growing rhizobia (under the genus *Brady rhizobium*) and (n) the fast growing groups (under the genus *Rhizobium*).

Why inoculation? Although, native rhizobia are present in the soil but all of them are not capable of forming nodules. Some of the strains are highly effective in this respect while others are partially or completely effective. It is also reported that native flora gradually lose their efficiency. Hence, the artificial inoculation with tested effective strains, should be taken up as a comparatively means, cheap insurance for obtaining optimum yield.

Carrier-based inoculants are usually coated for the introduction of bacterial strains into soil. However, alternative inoculation methods are necessary where seed treatment with fungicides and insecticides is needed or where seeds like groundnut and soyabean can be damaged when the inoculants is used with an adhesive. Direct contact with the acidic fertilizer can also be harmful for rhizobium.

Apart from application with seeds, the normal carrier-based inocula can also be separately applied. While all methods of application were successful under favourable conditions, "Liquid" and "Granular" methods were superior to seed inoculation under adverse conditions.

Azotobacter and *Azospirillum*

Many genera and species of N_2 fixing bacteria have been isolated from the rhizosphere of various cereals. Thus, mainly belong to *Azotobacter* and *Azospirillum* genera.

These are free living bacteria and fix atmospheric nitrogen in cereal crops without any symbiosis. They fix 15-20 kg/ha nitrogen per year. *Azotobacter* sp. also has ability to produce antifungal compounds against many plant pathogens. They also increase germination and vigour in young plants leading to an improve stand in crops. There is evidence in the literature that *Azotobacter* biologically control the nematode diseases of plants.

Phosphate Solubilizing Micro-organisms

Phosphorus is also one of the major element required for plant growth and higher yields. This element is necessary for the nodulation by *Rhizobium* and even to nitrogen fixers, *Azolla* and BGA. The phospho-micro-organism mainly bacteria and fungi make available insoluble phosphorus to the plants. The root fungus association or mycorrhiza has high potential in accumulating phosphorus in the plants. Mixture of charcole and soil is satisfactory material for these micro-organisms in order to prepare commercial inoculants. It is reported that microphos cultures increase yield upto 200-500 Kg/hectare and thus, 30 to 50 kg superphosphate can be saved.

Next to nitrogen, phosphorus is the vital nutrient for plants and micro-organisms. Superphosphate is one of the common forms of phosphate fertilizers. These fertilizers are prone to be fixed in the soil.

Rock phosphate is one of the basic raw materials for phosphatic fertilizers. Direct application of rock phosphate is limited to acidic soil, while in other types of soil the applied phosphate becomes insoluble within a short time. Monocalcium phosphate are converted to dicalcium phosphate which is slowly available to plants. Under such conditions large amount of phosphorus is fixed in the soil which is unavailable to the plants.

Several soil bacteria and a few species of fungi possess the ability to bring insoluble phosphate in soil into soluble forms by secreting organic acids. These acids lower the soil pH and bring about the dissolution of bound forms of phosphate.

Vesicular Arbuscular Mycorrhizae (VAM)

The symbiotic association between plant roots and fungal mycelia is termed as mycorrhiza (Fungal roots). These fungi are found associated with majority of agriculture crops. They are ubiquitous in geographic distribution occurring with plants growing in arctic, temperate and tropical regions alike. These fungi are obligate symbionts and have not been cultured on nutrient media.

VAM fungi infect and spread inside the root. They possess special structures known as vesicles and arbuscules. The arbuscules help in the transfer of nutrients from the fungus to the root system and the vesicles, which are saclike structures store P as phospholipids. VAM have been associated with increased plant growth, and with enhanced accumulation of plant nutrients, mainly P, Zn, Cu and S mainly through greater soil exploration by mycorrhizal hyphae. It has also been suggested that VAM stimulate plant growth by physiological effects other than the enhancement of nutrient uptake or by reducing the severity of diseases caused by the soil pathogens.

Azolla

Azolla is a floating fresh water fern inside which grows the nitrogen fixing BGA Anabaena. It contains 3.4 per cent nitrogen and also produces organic matter in soil. This biofertilizer is used for rice cultivation in different countries, such as Vietnam, China, Thailand, Philippines. It is also used in fish culture ponds these days. This can be easily grown in cooler regions. There is need to develop tolerant strains to high temperature salinity, and pests and disease resistance for its wider adoption. Field trial indicated that rice yields are increased by

0.5-2 t/ha due to *Azolla* application. In China, about 18 per cent increase has been observed. In fish culture ponds, application of 25 kg/ha per day *Azolla* biof ertilization has been worked out. Recent studies have revealed potentialities of *Azolla* as a nitrogenous fertilizer in carp culture ponds. Application of different doses of *Azolla* in fish culture ponds shown that a minimum of 25 kg N/ha/yr could be provided through application of 10-12 tonnes of AzoZ/a/ha/year. There are six species of *Azolla-A. caroliniana, A. nilotica, A. mexicana, A. filiculoides, A. microphylla* and *A. pinnata.* It grows in ditches and stagnant water. This fern usually forms a green mat over water. The plant has a floating, branched stem, deeply bilobed leaves and true roots which penetrate the body of water. The leaves are arranged alternately on the stem. Each leaf has a dorsal and ventral lobe. The dorsal fleshy lobe is exposed to air and contains chlorophyll. It has an algal symbiont *(Anabaena azolla)* within the central cavity. *Azolla* is readily decomposed to NH_4 which is available to the rice plants. It has also been observed that N contents of rice receiving *A. mexicana* are 10 and 35 kg/ha respectively higher than that in controls.

Blue Green Algae

Blue-green algae constitute an important group of micro-organism capable of nitrogen fixation. Most of the species possessing nitrogen fixation ability to the order Nostocales and Stigonematales. Over 100 species of blue-green algae are known to fix atmospheric nitrogen. These have been found to be very effective on the rice and banana plantation, In field condition, overall increase in the gram yield of rice is amounted to about 586 kg/ha. In case of crops other than rice, algalization increased nearly 34 per cent yield.

India is one of the countries where agro-chemical conditions appear to be favourable where blue-green algae technology has been put forward. In some parts of the country, production of BGA inoculants has been commercialised. Producing inoculum in artificially controlled conditions is well-defined, but relatively expensive. On the other hand open air soil culture is more simple, less expensive and easily adaptable by tile farmers. Field scale production of algae biofertilizer is also possible, 20-25 kg dry algae can be obtained on 40 m field. Adopting this method, 15 t/ha of /et BGA can be obtained by the farmers. Farmers can also produce algae for countryard of the house.

Plant Growth Promoting Rhizobacteria (PGPR)

A group of rhizosphere bacteria that exert a beneficial effect on plant growth is referred as PGPR. They belong to several genera, *e.g. Actinoplanes, Agrobacterium, Alcaligenes, Amorphosporangium, Arthrobacterm Azotobacter, Bacillus, Celhdomonas, Enterobacter, Erwinia, Flavobacterium, Pseudomonas, Khizobium* and *Bradyrhizobium, Streptomyces* and *Xanthomonas. Bacillus* spp. are appealing candidates for biocontrol because their endospores are tolerant to heat and desiccation. Currently *Pseudomonas* strains also suppress major plant pathogens. Take all, an important root disease of wheat caused by *Gaeumannomyces graminis* var. *tritici* is controlled by *fluorescent Pseudomonas.*

PGPR are thought to improve plant by colonizing the root system and pre-empting the establishment of suppressing deleterious rhizosphere micro-organisms on the roots. The plant growth promoting micro-organisms improved potato growth and yield in short-but not long-rotation soils, primarily by suppressing cyanide-producing deleterious rhizosphere micro-organisms. Large populations of bacteria established on planting material and roots become a partial sink for nutrients in the rhizosphere, thus reducing the amount of C and N available to stimulate spores of fungal pathogens or for subsequent colonization of the root.

In field trials with wheat, potato, sugarbeet and zinnia conducted showed significant yield increases varying from-7 to 136 per cent with an average increase of 7-35 per cent in different crops over the control. Seed treatment with *B. subtilis* increased yield of carrot by 48 per cent, oats by 33 per cent and groundnut upto 37 per cent and it has been marketed as a treatment for groundnut in USA. A multitude of factors could account for inconsistent results, given the complex interactions among host, inoculated organisms, other rhizosphere organisms and the environment which are discussed earlier. Research with PGPR in India is at a low level.

Status of Biofertilizer in India

As mentioned earlier the total potential of biofertilizers is very high and existing production units do not have the capacity to meet the requirement at the right time. Most of the production units do not have adequate instruments, production techniques, skilled staff, etc. for production of quality based inoculants. The result is the production of poor quality products which are directly responsible for the failure, so there is a need for upliftment of biofertilizer industry in India.

Biofertilizers differ from fertilizer in the sense that the former do not directly supply any nutrient to crop plant and are the cultures of some specific bacteria and fungi. Bacterial cultures like *Rhizobium, Azospirillum, Azotobacter* have the ability to fix atmospheric nitrogen which in turn increases nitrogen supply to crop. Bacterial cultures of *Pseudomonas* and *Bacillus* species and fungal cultures of *Pencillium* and *Aspergillus* species help to convert insoluble phosphates into soluble (plant soluble) phosphate and thus, improve phosphate availability to the plants. Similarly, fungi like vesicular arbuscular mycorrhizae (VAM) increased nutrient uptake particularly of P, Zn and other micronutrients due to increased contact of roots from larger soil volume. It is important to mention that so far biofertilizers available are able to take care of nitrogen only. Therefore, other plant nutrients have to be applied from other sources.

At present, in India there is a gap of about 10 million tonnes of plant nutrients between removal by crops and replenishment through fertilizers. Supply of nutrients from organic manure has not so far been able to fill up this gap and also in years to come nutrient supply from the source is unlikely to improve due to competitive demand for alternate uses like fuel and fodder. Of late, biofertilizers are being promoted as an important component in supplemei iting plant nutrient need of the country. It is a good beginning but we still have a long way to go.

India is one of the important countries in biofertilizer production and consumption in the world. The present production capacity of different biofertilizers production units in the country is about 4500 tonnes per annum. The maximum production capacity is in Agro Industries Corporation followed by State Agriculture Departments, National Biofertilizer Development Centre, State Agriculture Universities and private sector. Among the different states, the maximum production capacity is in Tamil Nadu followed by M.P., U.P., Gujarat, Maharashtra. The actual production of biofertilizer during 1994-95 was about 2000-2500 tonnes.

The installed capacity of BF production in different fertilizer industries in about 4001/annum. Main BF producing companies are GSFC, MLF, SPIC, etc. Some more fertilizer companies (IFFCO, KRIBHCO, NFL, RCF, etc.) are likely to start production soon and many more are planning to join the biofertilizer business.

Based on the area under different crops and dose of biofertilizer to be applied, the National Biofertilizer Development Centre (NBDC),

Ghaziabad and Biotech Consortium India Ltd. (BCIL) have estimated the total requirement of biofertilizers *(Rhizobium, Azotobacter, Azospirillum* and *Blue Green Algae)* to be about 5.07 and 3.44 lakh tonnes, respectively. The recent estimated potential demand of different kinds of biofertilizers by Government of Tami Nadu are *Rhizobium* 35 thousand tonnes; *Azospirillum* 482 thousand tonnes; *Azotobacter* 162.61 thousand tonnes: *Blue-Green Algae* 267.72 thousand tonnes, *Azolla* 20.38 thousand tonnes and phosphate solubiliser 275.51 thousand tonnes. The total of all these amounts to be 12.44 lakh tonnes which is significantly higher than the estimates of NBDC and BCIL, mainly because they did not indicate phosphate solubiliser and their estimates for *Azospirillum* were also low.

Production technology of biofertilizer is relatively simple and its installation cost is very low compared to chemical fertilizer plants. Organised promotional and marketing strategy for successful biofertilizer business is very important. Most of the biofertilizer units lack in this respect. Those who have very good organised marketing network, have done excellently well. For example, fertilizer company like GSFC has more than 200 farm information centres-cum-depots situated in remote areas. These are manned by experienced agricultural graduates to provide the technical know-how, conduct the demonstration and give before and after sales services to the farmers. In order to provide BF upto village level, GSFC has established its own distributor's network. MLF and SPIC have also well Response of biofertilizers have been obtained in cereals, millets, pulses, legume oilseed, sugarcane and cotton grown under different agro-climatic conditions. In general, increase in yield due to biofertilizer application was of the order of 15-20 per cent. It is found that biofertilizer like *Rhizobium* can supply 20-25 kg N/ha. Of late, complex biofertilizers have been reported to be more effective in increasing nutrient supply and crop yield.

To promote biofertilizer use, fertilizer industry is engaged in a number of activities like (i) field experiments on Research & Development (R&D) farm (ii) field experiments on farmers' fields (iii) farmers' meetings and field days (iv) seasonal campaign (v) farm youth training programme (vi) audio-visual display to educate farmers (vii) preparation of distribution of literature on biofertilizer in local languages (viii) distribution of paper stickers to institutional agencies, dealers, etc. (ix) publicity through local newspapers and (x) publicity through exhibitions, fairs, festivals, etc. Though there had been a lot of improvement in biofertilizer development and promotion in supplementing plant nutrient supply, the bottlenecks which still need to be sorted out are short shelf life, lack of suitable carrier material, suitable of high material, susceptibility to high temperature, problem in transportation and timely distribution in remote areas, etc. Besides standard packaging and quality control measures, etc. will help to develop healthy market in the time to come. There is urgent need to work out aggressive promotional and marketing strategies for gaining momentum to BF use in the country.

Table 3.7: Estimated Installed Capacity for Biofertilizers in Different States

State	Installed Capacity (t)
Tamil Nadu	1100
Madhya Pradesh	700
Gujarat	450
Maharashtra	250
Andhra Pradesh	175
Karnataka	40
Haryana	75
Orissa	75
West Bengal	72
Rajasthan	90
Punjab	75
Total	3102

Thrust in Research and Development

The rising cost of chemically fixed nitrogen fertilizer and the massive inputs of non renewable resource will be devastating. As an alternative there is urgency of major research effort in biological rorogen fixation. Nitrogen in the form of ammonia is used by plants and microbes as a building block : or the synthesis of amino acids and of other nitrogenous compounds. The conversion of atmospheric Nitrogen into ammonia by the nitrogen fixing microbes, mostly certain bacteria and blue green algae is called biological nitrogen fixation. Although leguminous plants have been used extensively in agriculture for centuries, it is possible to exploit these nitrogen fixing plants meaningfully. The use of new leguminous crops as well as the breeding of more vigorous and effective traditional leguminous crops can be helpful in food production. As the bacteria are the agents of fixation in bacterial plant association, plant breeding has been neglected to enhance the effectiveness of this association.

Rhizobial technology and soil microbiology are the neglected areas. It would be a right type technology that all farmers, could use Legume-Rhizobial combinations constitute, a built-in nitrogen source for food crop production. Approximately 75 per cent of the nitrogen utilised by legume comes from the soil, with only 25 per cent from BNR. It is necessary to increase the rate of nitrogen fixation achieved, which would have a multiplier effect on the enhancement of crop productivity. There have been many research frontiers in BNR The two genetic systems involved are micro-organisms and the higher plants and their greater potential for improving the both. In nature, certain nitrogen fixing agents do not perform well. As a result their nitrogen fixing efficiency remains less though their abundance is much. It is necessary to make them highly competitive through genetic engineering. It will help them to establish an association with the plants in competitions with the organisms occurring naturally in the ecosystem. Wherever possible, the agents should have a broad spectrum in their specificity so that they can associate with a variety of plants. Thus, the competitive strains developed could be introduced into the soil for better interaction of plant bacterial association.

Although nitrogen fixers are present in the soil, enrichment of soil with effective strains is much beneficial for the crop yields. Chemical fertilizers have temporary effect while biofertilizers have permanent effect without any production problem. Use of composite biofertilizers can increase soil fertility. The cost is too much low for the biofertilizers and its proper use does not only include a correct application of the inoculant to the seed-soil but also good crop management and provision of additional capital nutrients for-the plants and inoculum. Scientists have proved that the biofertilizers are cost effective, cheap and renewable source to supplement the chemical fertilizers. The rhizobia inoculant has been found to be effective for pulses and legume oilseeds like soyabean, groundnut and the blue green algae for low land paddy in our country. Much of the world today is concentrating on rhizobial technologies and the culture of other organisms having superior nitrogen fixing capabilities. As companion the legumes are being used as the soil improving crops and inter-crop or in rotation with cereal grains and other high energy crops as is being practiced in China, South America, Africa, Philippines and in India. They are becoming apart of cropping systems research in total farming systems. Active linkages of scientists engaged in basic fundamental research in biological nitrogen fixation with those in mission oriented, applied and problem solving research would speed up the process. Considering the prospects of biofertilizers in the country, the biofertilizer development centres are being established both in government and private sector. It is possible to establish joint venture in biofertilizer agro based industry.

❑❑❑

4

Nitrogen Fixing Micro-organisms : Symbiotic

Nitrogen is one of the most important essential plant nutrient to complete the life-cycle of plants, which is required in a huge amount by the plants. It is abundantly available in the atmosphere. Approximately 78 per cent of total atmospheric gases by volume and 75.5 per cent by weight is nitrogen. The large percentage of nitrogen is in the form of gas or molecular nitrogen (N_2) which is entirely unavailable to crop plants in such form. For making it available to plants, it is essential that this molecular nitrogen may be covered into combined nitrogen (NH_4^+). This nitrogen conversion phenomenon is termed as biological nitrogen fixation. The organisms responsible for such type of nitrogen fixation are known as "nitrogen fixers" or "Diazotrophs", *e.g. Rhizobium, Azotobacter, Azospirillwn, Clostridium.* Blue Green Algae, etc.

Nitrogen fixing organisms (NFOs) have very little in common with each other apart from being prokaryotic and diazotrophic. They may be symbiotic, non-symbiotic or form casual associations with other organisms; they may fix N under aerobic, anaerobic or microaerophilic conditions. Some organisms are capable of fixing N independently of other species, however, some can grow and fix N at a higher rate in the vicinity of other organisms which can provide them with their growth requirements. They are to be found in greater numbers in association with, for example, the rhizosphere of plants. These associations may vary from rather loose, non-specific attractions between plants and bacteria to those in which the plants seems to be able to encourage the growth of more restricted range of micro-organism in its vicinity. The associations are often specific. Some associations are such where both partners undergo considerable morphological and biochemical differentation in forming the N_2-fixing relationship.

Nitrogen is the most limiting nutrient in food production. The biological nitrogen cycle is responsible for a turnover of 10^8-$10^9$1N per hectare on the earth in which biologically fixed N is one of the inputs. Non-symbiotic and associative N fixation is considered to occur at magnitudes that may be of agronomic significance. The apparent potential for biological nitrogen fixation (BNF) associated with cereals exceeds its present utilization.

Since Winogradsky, a German microbiologist in 1893 established that *Clostridium pastaurianum* which fix atmospheric N and Beljerinck, a Denmark microbiologist described the first *Azotobacter,* the list of N-f ixers is increased gradually. Many different genera and strains of N_2-fixing bacteria can be isolated from the soil and the roots. A list of organisms which fix nitrogen in soil has been given. Since in agriculture few organisms are studied for their N-fixing qualities from the list, hence, detail discussions was confined to very few of them (Table 4.1).

Table 4.1: Important Nitrogen Fixing Organisms

Nature/Organism (s)	Remarks
Anaerobic	
Clostridum pasteurianum C. butyricum	Isolated from soil, fresh water and strictly anaerobic; contribute about 33 kg nitrogen/ yr/ha in rice field;
Methanosarcina barken	Strictly anaerobic but the contribution to N-fixation is unknown
Methanococcus thermolithotrophicus	
Anaerobic (Partial or Microaerobic)	
Enterobacter cloaceae	Isolated from soil and water; partial aerobic
Klebsiella pneumoniae	Isolated from water and soil and roots of grasses
Aerobic	
Azotobacter vinelandii A. paspali	Isolated from soil, leaf and roots of plants; fixes atmospheric N- as free living bacteria;
Beijerinckia derxii	All strictly aerobic but some species can tolerate low O_z
Photosynthetic Bacteria	
Rhodospirillum rubrum	Aerobic and aquatic
Symbiotic Diazotrophs	
Rhizobium spp.	Legumes
Bradyrhizobium japonicum Frankia sp.	Non-legumes

Biological Nitrogen Fixation

Biological nitrogen fixation involves the reduction of nitrogen to ammonia. This becomes incorporated into the organic nitrogen of the N-fixing organisms. Some free-living diazotrophs excrete NH_4 or amino acids into soil but the greatest contribution to the ammonia content of the soil comes from the death and subsequent purification of plant and animal material. Ammonia in solution exists as the NH_4^+ ion which remain in the upper layer of soil and is readily available for utilization by plants and micro-organism. This utilization may take the form of association, but in competition with association is the process of nitrification in which some bacteria oxidise NH_4^+ to nitrite and nitrate. The oxidation of NH_4^+ to nitrite is mediated by the bacterium *Nitrosomonas* and nitrite to nitrate by *Nitrobader*. Both bacteria are obligate aerobes. Nitrite in the soil is less tightly bound than ammonia and there, it is subjected to run off. Nevertheless, nitrite is more easily absorbed than ammonia by, plants from the soil, however, nitrate which is absorbed by the roots of plants, must first be reduced to ammonia before it can be incorporated into the organic compounds that the plant requires. This assimilarity reduction of nitrate (which can occur both in the roots and the leaves) occurs in two main stages and involves two enzymes, nitrate reductase and nitrite reductase (Fig. 4.1).

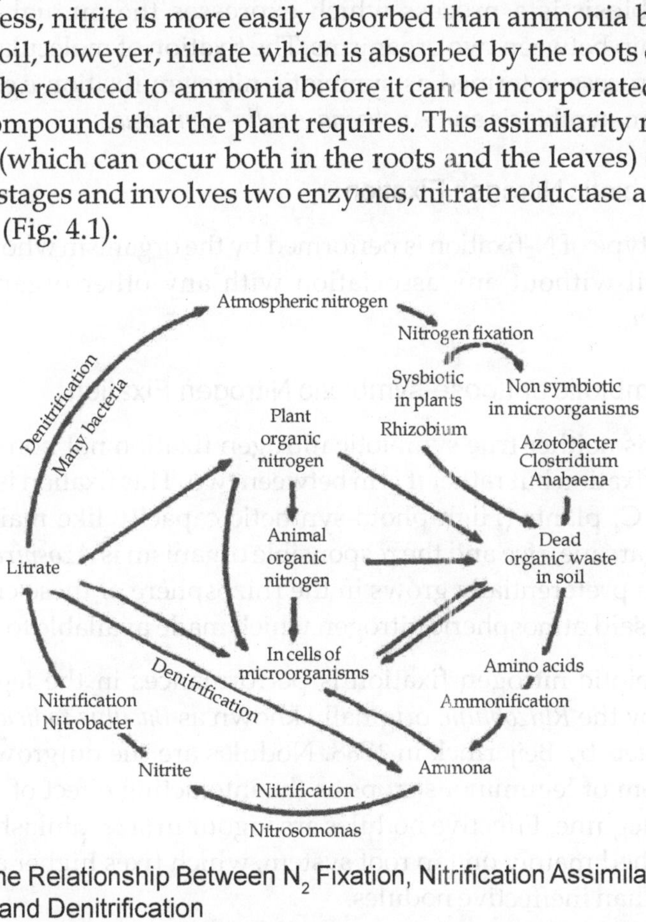

Fig. 4.1. The Relationship Between N_2 Fixation, Nitrification Assimilatory Nitrite Reduction and Denitrification

The accurate estimates of annual turnover of nitrogen in the biosphere vary from 100 to 200 million tonnes. The ratio between chemically fixed N and biologically fixed N ranges approximately from 1:4 to 1:2.5 and within biological fixation, the legume fixation is equivalent to or at least *'A* that of industrial fixation. The amounts of kg N fixed/ha/yr for individual legumes has been estimated is: 125-335 for alfaalfa, 85-190 for red clover, 80-150 for peas, 65-115 for soyabean, 65-130 for cowpea and 95-155 for vetch.

Types of Biological Nitrogen Fixation

The following are the three types of Biological-Nitrogen-Fixation:

Symbiotic Nitrogen Fixation

Symbiosis is a process which expresses the mutual beneficial partnership between two organisms. The fixation of molecular nitrogen by this process is termed as symbiotic nitrogen fixation *e.g.* Legume-Rhizobium-symbiosis and *Anabaena-Azolla-symbiosia.*

Non-Symbiotic Nitrogen Fixation

This type of N-fixation is performed by the organism who live freely in the soil without any association with any other organisms *e.g., Azotobacter.*

Semi-Symbiotic or Loose-symbiotic Nitrogen Fixation

This is neither true symbiotic nitrogen fixation not non-symbiotic nitrogen fixation but rather it is in between two. This fixation is generally found in C_4 plants (High-photo-synthetic capacity like maize: jowar, bajra, sugarcane, rice and the responsible organism is *Azospirillum.* This bacterium preferentially grows in the rhizosphere of these crop plants and fixes said atmospheric nitrogen which made available to plants.

Symbiotic nitrogen fixation is performances in the legume root nodules by the *Rhizobium,* originally known as *Bacillus radicicola* which was isolated by Beijerinck in 1888. Nodules are the outgrowth on the root system of leguminous crops as the interacting effect of *Rhizobium* and host legume. Effective nodules are vigour in size, pinkish in colour and attached mainly on tap root system, which fixes higher amount of nitrogen than ineffective nodules.

Factors Affecting Nitrogen Fixation

Inorganic Nitrogen

Low concentration of nitrogen as NH_4^+ or $N0_3$-to the extent of 10-15 kg N ha^{-1} increases the nodulation, however, higher concentration of N reduces nodulation and N-fixation.

Soil pH

Acidic condition is deterious for legumes and also for colonization of *Rhizobia.* Soil acidity also results in the deficiency of Ca, Mg, K and Mo which indirectly effect nodulation. Toxicity of Fe and AI in acidic condition is responsible for lower N-fixation. Detrimental effect of acidity can be improved by liming the acidic soils and pelleting the inoculated seeds. The optimum range for better nodulation and N-fixation is neutrality to slight alkalinity. In high salt content soil, growth of legumes and N-fixation become reduced.

Temperature

The most congineal temperature for bacteriod tissue has been reported from 20-30°. The temperature above and below this range depressed the nodulation and N_2-fixation.

Light

Light through photosynthesis affects the nodulation and N-fixation. Sixteen hours light in the temperature range of 15-20°C has been reported better for N-fixation under controlled condition.

Growth Substances

Different types of growth substances affect nodulation and N-fixation differentially. Lower levels of IAA conducive for early nodulation while higher concentrations of NAA inhibit nodulation.

Use of Biocides

It is generally observed that all biocides (fungicides, weedicides, insecticides, etc.) at the recommended dose do not adversely affect nodulation. However, persistence of these chemicals in soils at the high concentration may reduce the nodulation. DDT at lower concentration of 1-10 ppm has been reported to stimulate nodulation.

Genus-Rhizobium

The genus Rhizobium was established in 1889 by Frank based on its ability to form nodule on the root of legume. The present classification of *Rhizobium* depends mainly on the work of Fred and Coworkers.

The genus *Rhizobium* alongwith the genera *Agrobacterium* and *Chroniabacteriam* comprise the family Rhizobiaceae and divided into three main groups: *AzorhizobUmt* produces stem nodulation in plant *Sesbania rostrata;* the second group is Bradyrhizobium with two well-defined species

B. japonicum and *Brudirhizobium* sp. produce nodules in the roots of soyabean, lupin, cowpea, green gram, red gram, chickpea and groundnut. The members of this groups are slow growing strains producing alkali on Yeast Manitol Agar (YMA). The third group is *Rhizobium* establishing great relationship in pea, lentil, bean, lathyrus, berseem and lotus. They are classified as fast growing acid producing on the Yeast Manitol Agar (YMA), the important species are :

R. phaseoli, R. trifolii, R. legiiminosarum and R. *meliloti.* It is opined that the DNA-base ratios of fast growing species. The lack of homology between the DNA of *B. japonicum* and DNA of representative strain of *R. leguminosarum* is sufficient to justify the separation of *Rhizobia* into two genera.

The *Rhizobia* are gram negative, unicellular cell size less than 2 micron wide and short medium rod, pleomorphic, motile with peritrichous flagella and accumulate poly *p*-hydroxy butyrate granules. The physiological studies indicated that the bacterium uses monosaccharides and disaccharides as carbon sources which are supplied by legumes through photosynthates. They are aerobic and the growth media is YMA.

Rhizobia

Significance role of *Rhizobium* in fixing atmospheric N and hence, enhancing soil fertility, has been well established since time immemorial, however, commercialisation of their use is a relatively new venture. In view of their apparent low cost and risk free application leading to yield advantage both for the current crop as well as for the subsequent crop grown on residual fertility, their large scale application in intensive farming calls for systematic efforts. Their use is especially important in view of the fact that in the foreseeable future it may not be at all possible for the country to be self-sufficient in the production of nitrogenous and phosphatic; fertilizers. *Rhizobium*-a symbiotic bacterium, colonises on the roots of specific legumes to form root nodules which fixes atmospheric nitrogen. In a way, the nodules act as mini factories for ammonia production, *Rhizobium* has been widely studied and extensively used as a biofertilizer because of its proven ability to fix and contribute N for the benefit of a leguminous host plant. In view of crop specificity of a *Rhizobium,* selection of an appropriate strain as an inoculant for a given host genotype is of utmost importance for desirable results. Dawdle and Bohlool suggested a simple strategy to develop an improved practical inocuiant. The various steps of this strategy are (i) use of a dominant, competitive indigenous strain specific to a particular geographic region which is adapted due to natural selection, (ii) isolation of selected indigenous strain, its addition to the seed in high number and its subsequent use in the same region under favourable environmental conditions, and (iii) mutants with increased effectivity may be isolated from the indigenous strains which remains as competitive as the wild type.

The *Rhizobium* legume association could fix upto 100-300 kg N/ ha in one crop season and in certain situation leave substantial N for the following crop. The N advantage through this symbiosis is greater in soils poor in available N. The response of pulses inoculation with *Rhizobium* varies widely but, by and large, has been found to be positive. Field experiments conducted at several locations and on different pulses, revealed yield increases ranging between 9 to 16 per cent. The response of *Rhizobium* is also believed not only to be crop specific but cultivar specific also. In field experiments carried out on groundnut at ICRISAT, Hyderabad, *Rhizobium* strain NC-92 was found to give best results only with the groundnut cultivar Robot. These findings stress the need to isolate and multiply cultivar specific elite *Rhizobium* strains suitable for different crop varieties, for obtaining enhanced productivity. Also, response to *Rhizobial* application can vary widely depending upon the soil *Rhizobial* status as well as *Rhizobial* strain. In soils with higher than 1000 *Rhizobial* gm, the response is minimum, while in soils with less than 100 *Rhizobial,* the response could be extremely high.

Rhizobium/Legume Symbiosis

Legumes are belong to family leguminosae with subfamilies *Caesalpinioodeae, Mimosoideae, Papilionoideae* and *Sewartziozeae.* The extents of nodulation in the family is not known exactly, but is least in *Caesalpinioodeae,* where only about 30 per cent of the species are nodulated and greatest in the *Papilionoideae.* With certain genera, for example, Cassia, not all species may be nodulated, possibly, reflecting heterogeneity in the genus.

Nodulation does not occur in the field until the first leaves are unfolding and N-fixation does not normally commence until the plant can safely divert a proportion of its photosynthate to this end. This requires synchronisation of plant and *Rhizobial* development at a number of stages, begin in the soil.

Seeds of legumes do not usually carry their own indigenous population of potentially symbiotic bacteria, is well established. This indicates that infection by *Rhizobinm* after germination of the host is dependent upon the availability of suitable *Rhizobia.* Earlier it was thought that N-fixation can only occurs in symbiotic association with legumes and N-fixation by free living *Rhizobia* was, therefore, not possible. It is reported *Bradyrhizobium japonicnm* can fix nitrogen undermicroaerobic conditions. The various stages of *Rhizobinm* legume symbiosis is discussed.

Methods for Study of Legume Root Nodulation

Isolation

Wash the root system of the test legume plant in water, a well-formated; healthy pinkish nodule on the tap root is carefully cut out with a portion of the root attached to the nodule. The nodule is surface sterilized for 5 minutes in 0.1 per cent mercuric chloride in water and repeatedly washed with sterile water to remove the chemical. The nodule is then washed in 70 per cent ethyl alcohol for 3 minutes followed by more washings with sterile water. The nodule is now crushed with a sterile glass rod in a small aliquot of sterile water. Serial dilutions of the suspension are then made so as to obtain sparse and distinct colonies, when an aliquot of the appropriate dilutions is plated on YEMA medium. After six to seven days of incubation, distinct colonies of bacteria are picked up and transferred to agar slants for further identification. Several strains of *Rhizobiuym* of varying effectiveness, often associated with a related genus *Agrobacterium,* are isolated from the same plant. Obviously, further screening of such isolated becomes an essential step.

Differentiation of *Rhizobium* from its Common Associate *Agrobacterium*

Congo Red Medium

In Congo red incorporated agar medium (2.5 ml of a 1 per cent solution per litre of yeast extract manitol agar), rhizobia stand out as white, translucent, glistening, elevated and comparatively smaller colonies with entire margins in contrast to stained colonies of *Agrobacteria* concentration. Hofer's alkaline medium is other medium for differentiating these two bacteria. *Agrobacteria* tolerate higher pH while *rhizobia* is unable to do so. Therefore, growing bacterial isolates on an alkaline medium would serve as a useful criterion for distinguishing the two allied genera. *Lactose agar* is another medium for differentiation. *Agrobacteria* utilize lactose to form a reduced product, ketolactose, through the enzyme ketolactose. When Benedict's reagent is poured over agar medium containing lactose on which nodule bacteria are growing, the formation of yellow colouration due to Cu_20 indicates the presence of agrobacteria.

Tests for Nodulation

Rhizobium isolates with a range of effectiveness, occurrence of non-nodulating and also pathogenic strains among the isolates can be obtained. Therefore, testing strains for their ability to nodulate and for their effectiveness under sterile conditions is of prime-importance.

Small seeded legumes are grown on Agar-agar, vermiculite or sand. Any nitrogen-free plant nutrient is used to make agar slants. Seeds are surface sterilized 0.1 per cent silver nitrate and sown on the substrate in tubes. On germination, they are inoculated with a suspension of the bacterial isolate under test prepared in a small aliquot of the nutrient solution. The moisture content of the tubes should be checked and the nutrient solution replenished when necessary. Experiments which may run for months, the plants are examined for nodulation. A criterion for distinguishing the effectiveness is to record the relative dry weight of nodulated plants.

For larger seeded legumes isolated are tested for their efficiency by growing the inoculated legume in sterilized sand with N-free plant nutrient solution. The data on the appearance of plants in terms of colour and vigour, the number of nodules formed on the root system, dry weight and nitrogen content of the plants are collected. Temperature and light conditions be kept at optimum levels favourable for maximum performance of legume *Rhizobium* symbiosis. A standard assembly consisting of a bottomless screw capped bottle and a glass jar designed to permit automatic and continuous water-flow to the growing root system can be used for the routine testing of isolates of *Rhizobium* for efficiency in symbiosis. Alternatively, sterilized sand or soil in earthenware or porcelain pots can also be used to grow large.

Infection Test

Seedlings may have to be, examined for the percentage of infection for virulence and improving nodulation. This is carried out by growing plants on agar slants containing N free medium and inoculating them with desired strains. Seedlings are fixed in 4 per cent formalin and examined for infection threads in root hairs and emergence of nodule primordial on roots. Depending on the extent of infection, isolates may be rated for virulence. A simple device for continuous observation of the progress of infection has been designed by Subba Rao.

Nodulation

This growing tip of a 3-day old seedling (grown aseptically) is cut and the blunt end implanted in a small vial containing organic constituents (glycine, B-vitamins and sucrose) in solidified agar. The tip of the excised root is suspended in nitrate-free mineral salt medium in a petri dish or in a test tube into which a suspension of *Rhizobium* is added. In this way, the organic constituents are provided to the growing root through the cut end and the growing end of the root receives mineral constituents along with *Rhizobium* culture. Nodules develop after 2-3 weeks when lateral roots are fairly well developed.

Tissue and Cell Cultures

Tissue cultures in plant morphogenetic studies are well known. Explants meristem can be grown into an callus mass on a suitable medium. New plantlets are transplanted to soil, they grow into mature plants bearing flowers and fruits. Tissue culture techniques have extensive use in development of resistant varieties. In recent past, root calli have been used to shady the interactions of plant tissues with *Rhizobium* or *Azotobacter*. A mass of undifferentiated callus or a Suspension of isolated cells is inoculated with nitrogen fixing bacteria, such as *Rhizobium* or *Azotobacter* and incubated under controlled conditions. Samples of callus or cells could also be cut into thin sections and observed under light or electron microscope.

Studies have demonstrated that (i) Infection thread-like structures and bacteria or bacteriod containing cells could be seen in callus tissues or isolated cells inoculated with *Rhizobium;* (ii) Definite nirogenase activity could be detected in inoculated calli or cells indicating the establishment of symbiosis in such plant cell-bacteria associations; (iii) When cells of *Azotobacter* and carrot were grown on N-free media, not only the bacterial cells entered isolated cells of carrot but such cells showed nitrogenous activity; and (iv) Such studies are extended further, the day may not be far off when 'nitrogen fixing', plants are raised as clones through plartlets re-generated from bacteria incorporated cells of tissues.

Acetylene Reduction Assays

The enzyme nitrogenase, in addition to reducing nitrogen to ammonia can reduce certain other compounds like acetylene. This assay, is based on the nitrogenase-catalyzed reduction of acetylene to ethylene. It is simple, sensitive and inexpensive. The advantage of the technique includes facility for large number of *in situ* assays, economy, simplicity, effective gas exchange prior to incubation, easy removal of the gaseous end products and sensitive analysis with a flame ionization detection system after separation by gas chromatography. The technique is 10^3 times more sensitive than ^{15}N tracer technique. Samples are incubated in serum vials with purified acetylene at temperature, incubation period and illumination. After incubation, the gas from the container is removed for ethylene analysis by gas chromatographic (GC) fractionation followed by detection by flame ionization. After correction for the small amount of acetylene present as an impurity in ethylene the amount of nitrogen fixed is determined by the extent of reduction of acetylene to ethylene. Moles of nitrogen fixed is obtained by division of moles of acetylene reduced to ethylene by a factor of 6 electrons for reduction of nitrogen and 2 for acetylene.

Use of ^{15}N to Measure Biological Nitrogen Fixation

While ARA offers a rapid though indirect way of measuring nitrogenous activity and thus, the ability of a biological system to fix atmospheric nitrogen, any critical evaluation of results must pass the ^{15}N test. The technique formulated involves the use of ^{15}N as a tracer. The amount of nitrogen fixed by a system is determined as the percentage increase of the tracer in samples incubated with the labelled gas. An enrichment of 0.015 or greater atom per cent ^{15}N in the sample (expressed as 0.015 atom per cent excess) is usually taken as an extensively in the analysis of nitrogen fixation in whole cells and in cell-free extracts.

Multiplication of Rhizobia: Root Hair Curling

On germination of legume seeds they release chemical substances which attracts nature or introduced *rhizobia;* the chemicals include amino acid, homoserine, which stimulate *Rhizobium leguminosarum* growth. The study was conducted with pea seeds. The tap root of pea does favour *rhizobial* multiplication. However, as soon as lateral roots begin to emerge; release various chemicals which favours *rhizobium.* The first nodules, is common with those of most similar legumes, are formed on the tap root. Once a desired population of *rhizobia* has been built up, these can begin to influence and possibly infect the host plant. The first symptom is deformation of root hairs. Three types of deformation has been described: (i) branching (ii) moderate curling (iii) marketed curling. Not all hairs on a root respond and sometimes growth right through a *rhizobial* colony can occur without turning a hair. There are several theories proposed on root hair curling. The suggestion that auxin (Indole acetic acid) is an essential component. The curling may be due dialysable heat stable products of *Rhizobia* may be responsible. A curl could delimite an area when some vital substance accumulate. It is also possible that entry holes made in the crook of a curl may be more easily plugged to prevent entry of pathogens. Root nodules contain several bacteria other than *Rhizobia* but not pathogenic species. Before entry can be effected, *Rhizobium* of the right type must be correctly aligned on the hair. At this stage, *Rhizobia* are frequently seen to be embedded in mucilage, a material of legume origin, which may cover root to soil and also provide food for bacterial growth. The *Rhizobial* and host interactions was compared to antigen and antibodies in animals and it is predicted that some specific substance is produced which is identified as lectin and it is believed that this can contribute to legume *Rhizobial* specificity.

Lectins of host are responsible for binding to root hair surface and rhizobium. The binding sites on the parts are structurally and antigenically similar polysaccharides.

Formation of Infection Threads

The aligning of desired *Rhizobia* on root hairs is the first steps of complex interactions; the next step is the entry of *Rhizobium* into host cells. When desired bacterium firmly attached to the root hair a retractile spot becomes visible in the root hair wall, this indicates structural changes. After this infection threads become visible. This consists of one or more roots of bacteria aligned within a sheath of host cell wall origin, it grows down the root hairs. The infection thread grows from the root hair cell between and through cells of the root cortex often into considerable branching.

Nodule Formation

Nodules developments in the roots start once the cortical cells have been infected by *Rhizobia*. The development is of two types:

1. Those in which infection of new cells occurs by means of infection threads which are long and highly branched continuing infection of newly formed cells is depended upon the spread of infection threads. The nodules formed in this way are elongated, branched and designated as interminate; these are found in peas and clover and typical of temperate region.
2. In this type infection thread may not appear (peanut) and branched and ramify extensively (soybean). In this, nodules are more or less spherical in shape and are limited in size. Such nodules are determinated.

After the infection thread passes through the outer layers of cortical cells, the cells of the inner root cortex show increased RNA contents and begin to divide. Cell divisions are generally opposite protoxylene points in the root stele. The cell division occurs due to the presence of auxin and cytokinins together with some as yet unidentified factor. The advancing infection thread travels inner cortical cells, endodermis and possible the pericycle, the cells continue to divide in various planes and become infected. *Rhizobia* are released from the confines of the thread wall and develop into N-fixing bacteroids. The host cells lose their meristematic activity, the infection thread branches and infects the growing cells cut off by the nodule meristem. A continuous sequence of newly infected cells is formed. This process continues until long after the nodule has emerged from its subtending root. The outer layers of the nodule do not become infected; instead they form the cortex, within which form vascular traces, which connect to the state of the subtending root. There is a nodule endodermis continuous with the root endodermis and additionally, an endodermis around each vascular trace.

The nodules have all the bacteroid containing cells formed from cells individually infected by a branch of the original infection thread. According to Dr. Newcomb of UK, when an infection thread penetrates a cell it contains bacteria embedded in a matrix, mucopolysaccharide of bacterial origin. This in turn is surrounded by plasmalemma continuous with that of the infected cell. At the top of infection thread, the matrix appears to break through the host wall, releasing bacteria. In turn the bacteria break free from their own matrix and become individually surrounded by membrane envelops in the host cell cytoplasma, where they enlarge and differentiate into the N-fixing, bacteroids. The bacteria when differentiating into the N-fixing bacteroids show increasing invagination of the surface membrane and loss of mesosomes. The cell wall altered in structure and much permeable, largely due to alteration, and make it difficult to separate bacteroids. Changes in bacterial cell wall structure may be under host cell control. The outer wall may disappear and leave protoplasts. Such cell may lose viability. A relation between N- fixing activity and the presence of haemoglobin has been found. Haemoglobin is rare in the plant Kingdom; the two components appear to be coded by a different symbionts; haem by the *Rhizobium* and apoprotein by the host. The exact location of haemoglobin is still a matter of dispute but evidence is accumulating that it may be in the host cytoplasm.

According to Dr. J.I. Sprent (UK) the life of nodule is limited. The infected cells degenerate, turn greenish brown in the process. This has been attributed to the breakdown of haemoglobin into bile pigments. The studies also indicated that membrane envelops break down and cell contain degenerate as ageing proceeds. The cells may become invaded by free living *Rhizobia* which may later be killed by the action of phytolysosomes. Besides N-fixation, an efficient gaseous exchange system is clearly available. The pathways for aeration have been described in nodules of several crops. Proceeding from the soil, via lenticals on the surface to intercellular spaces in the centre of the nodule. Lenticals seem to be a common feature of the surfaces of determinate nodules, although the development is totally depend on water supply. Aerobic condition is necessary for N-fixation but in anaerobic condition N-fixation has been observed in soyabean.

In effective nodules are white, since they do not contain haemoglobin. Actually they are pasity on the host. Riboflavin is responsible for ineffective nodule. The development and effectiveness of nodule is under genetic control between host and *Rhizobium.*

In legume roots *Rhizobia* are incorporated in cells to form the N-fixing root nodules. This is extremely well balanced symbiotic system in which the host supplies the microsymbiont with all its carbon compounds and receives NH_3 in exchange leaving the *Rhizobia* free to concentrate on fixing N. An efficient O_z supply to the active bacteroids is required. This is achieved with the help of leghaemoglobin cytochrome P450 and by branch respiratory pathway.

Cross Inoculation Group

Cross inoculation group refers to a group of related legumes which can be nodulated by a particular species of *Rhizobium* isolated from any member of that group. Legumes has a definite preference for some specific bacterium and in turn the bacterium has preference for certain legumes. Later it became clearly apparent that there exists groups of legumes, each of which were nodulated by a different type of *Rhizobium.* Each of these groups of legumes and their specific *Rhizobia* constituted a cross-inoculation group and it generally held (i) that with each cross inoculation group would nodulate all other members of that group, and (ii) that *Rhizobia* isolated from one plant is cross inoculation group would not nodulate plants from other groups. Such observations develop the concept that plants mutually susceptible to nodulation by a specific *Rhizobia* constituted a cross inoculation group. This response of nodulation is fixed character and could be used as a parameter of classification of the bacterium (Table 4.2). This has been agreed upon that most of the legumes in India are infected by cowpea group except that of chickpea which behaved differently, thus indicate chickpea rhizobias is from different group. Bergey's manual of Systematic Bacteriology is 1984 proposed a different classification (Table 4.3). Because the cross inoculation grouping system is not perfect, since the bacterium has often been found to cross-infect between groups. Such type of behaviour has been found what is known as promiscuity in groundnut, cowpea and pigeonpea. It has been observed that *Rhizobia* isolated from soyabean can nodulate cowpea but not the vice versa. Nodulation of pigeonpea by *Rhizobium japonicum* has been reported by Singh.

Table 4.2 : Cross inoculation groups of *Rhizobium*

Rhizobium Species	Cross Inoculation Group	*Host*
R. trifolii	Clover group	*Trifolium*
R. meliloti	Alfaalfa group	*Melilotus, Medicago*
R. phaseoli	Bean group	*Phaseolus*
R. lupini	Lupine group	*Lupinus, Ornithopus*
R. leguminosarum	Pea group	*Pisum, Vicia, Lens*
*R. japonicum**	Soyabean group	*Glycine*
Rhizobium sp.	Cowpea group	*Vigna, Arachis*

*Slow growing rhizobia

Table 4.3 : New system of classification of root nodule bacteria

Bacterium	*Biovars*	*Host*
Robium leguminosarum	(a) *Viciae*	*Vicia*
	(b) *Trifolii*	*Trifolium*
	(c) *Phaseoli*	*Phaseolus*
Rhizobium meliloti	-	*Medicago*
Rhizobium loti	-	*Lotus*
Rhizobium friedii	-	*Glycine*
Bradyrhizobium		
B. japonicum	-	*Glycine*
Bradyrhizobium sp.	-	-

Legume plants of one or more genera or species which develop root nodules in association with the same varieties or species of *Rhizobium:* example: under per group *R. legwninosaram* can nodulate pea, lentil and lathyrus. However, Nobbe demonstrated with cultures that *R. lupini* did not nodulated with the pea but would do so with the lupin organism, and that beans and pea did not nodulated with isolated from Lupinus or Robinia. Atkinson found that the organisms from *Vicia sativa* and Dolichos bean would not cross infect. Further Hiltner demonstrated that nodule bacteria from *Trifolium pratense* would not nodulate *Medicago sativa.* These observation lead to the theory that separate *Rhizobium* species only nodulate specific related leguminous plants. Thus, Fried recognised seven groups, *viz.:*

1. *Rhizobium japonicum:* Also known an *Bradyrhizobium japonicum;* the bacterium produces nodules in soyabean, *Lupinus* sp. and cowpea miscellany. Bacteroids in nodules are longer than normal cells, slender and with only occasionally branched and swollen forms. They are slow growing bacteria (Soybean grain).

2. This is fast growing acid forming species of the pea group which mainly forms nodule in pea, broad beans, lentils and vetch. Bacteroids in nodules are commonly irregular with X, Y, star and club shaped form (Pea grain).
3. *Rhizobium lupini:* This species cause nodule formation on *Lupinoas* sp. and *Ornithopus* sp. Bacteroids are vacuolated rods and seldom branched (Lupin grain).
4. *Rhizobium meliloti:* The bacterium causes nodule formation in sweet clover, alfalfa and fenugreek; they are rod shaped, branched and grows fast on yeast extract medium (Alfaalfa gr.)
5. *Rhizobium phaseoli:* The bacterium produces nodules on temperate species of Phaseolus. Bacteroids are rod shaped and often vaculated with few branched forms (Bean gr).
6. *Rhizobium trifolium: Trifolium* spp. nodules are produced by this bacterium. Bacteroids are pear swollen, vacuolated, rarely X and Y shaped (Clover gr).
7. *Rhizobium* sp.: Nodulate cowpea group. Bacteroids are irregular shaped (Cowpea gr).

Fungicide Enhancement of Nitrogen Fixation: Current practices used for legume production include inoculation of seeds with *rhizobia* and treatment of the seed with fungicides to reduce seed rot and seedlings damping-off resulting from infection by soil-borne and soil-dwellers pathogens. Since the fungicides are biocides and are not specific for the pathogens, they also inhibit the soil biota *viz.* fungi, bacteria and other groups of micro-organisms. Thus, seed protection and seed inoculation are frequently incompatible. Such in, compatibility are particularly serious when the fungicides are applied to seed of legumes, because the consequent suppression of species of *Rhizobium* may result in nodulation failures of reduction in quantities of N_2 fixed by *legume-Rhizobium* symbiosis. Besides this chemofobia (fear of chemicals) which is prevailing among cultivators that pesticides have toxic effects on legume nodulation is justifiable. Reasons for decline in *rhizobial* activity in relation to micro-organism.

Colonization of the rhizosphere by *rhizobia* is probably important in determining extent of nodulation of legumes. This necessitate that *rhizobia* should proliferate as a first step for successful nodulation and this has been stressed by several microbiologists. A correlation has been established between failure to nodulate and low *rhizobial* count in the rhizosphere; there are several other reasons which affect *rhizobial* population and colonization of plant.

There are several ways to overcome these factors like using booster dose, increasing the number of *rhizobial* propagules and use of fungicides, singly or simultaneously.

Experimental evidence indicated that the population of *rhizobial* increased rapidly around germinating seeds and developing roots of soybeans, but the numbers of these bacteria fell abruptly. The decline was delayed by treatment of the seeds with benomyl so that, as the root system grew and opportunities for *rhizobial* infection of root hair arose, the population of *rhizobia* was maintained at higher levels than on plants not receiving the fungicide. The beneficial effect of benomyl was known and evident on first day. Such a stimulation was not restricted to fungicide treated seeds because a similar beneficial influence was noted by soil amendment with fungicides/ antibiotics. This enhanced colonization associated with the use of fungicides and a fungicide resistant *rhizobia* strain may be account for greater yield and large number of nodules on the several plants receiving the fungicide.

Effect of benomyl and the antibiotics on colonization by *rhizobia* probably results from their effects on micro-organisms antagonistic to the *Rhizobium*. Microbial antagonisms towards *rhizobia* and their symbiotic performance has been demonstrated by several studies: *Trichoderma viride* reduced the *rhizobial* population in non-sterile soil. Other organisms responsible are *Bdellovibrio* spp., bacteriophages, lytic micro-organisms and myxobacteria. These are potentially harmful to the *rhizobia*. The enhanced development of *Rhizobium japonicum* in the *rhizosphere* after the addition of streptomycin and erythormycin to soil suggests that prokaryotes are important in limiting the growth of *Rhizobium* spp. the lesser number of protozoa after addition of these prokaryotic inhibitors probably result from the lesser number of bacterial prey than a toxicity to the predators.

Development of resistance of *Rhizobium* towards fungicides used for the seed treatment and soil application, at a concentration for which they are used to control is desirable approach to enhance *rhizobial* colonization in the *rhizosphere*. Several studies indicated that the tolerance or resistance towards benomyl, thiram and mancozeb in *Rhizobium meliloti* (alfalafa strain), *R. phaseoli* (Bean strain), *Rhizobium* (Cowpea). *R. trifoli* (red clover) and *R. japanicum* (Soyabean) has been developed automatically, if these bacteria are continuously grown in the medium containing these fungicides. It is concluded that tolerant strains can detoxify these protectants, it is possible that the bacterial inoculants and the fungicides can be applied to legume seeds together without the letter destroying the former.

In a study it was observed that application of thiram to seeds inoculated with a thiram resistant strain of *Rhizobium* resulted in significant increase in dry weight and nitrogen contents of the plants compared with inoculation or thiram treatment alone. The young plants from such treatment were found to bear more nodules than untreated ones. Nodulated beans plant begins to fix nitrogen about 4 weeks after planting, so that an increase in number of nodules containing an effective strain of *R. phaseoli* could result in greater amounts of nitrogen being available to the plants treated with thiram during early part of growth. Inoculation of thiram treated seeds with a thiram-resistant strain of *R. phaseoli* resulted in higher numbers of *Rhizobium* in spermosphere and *Rhizosphere* of kidney beans as compared with seedlings that arose from untreated seeds.

The tolerance or resistance was developed in *Rhizobium* strains because the detoxified, the protectants it is possible that the bacterial inoculants and fungicides can be applied to legume seeds together without the letter destroying the former. Hence, the fungistatic agents might be employed to control fungal pathogens without jeopardizing the potential capacity of the *rhizobia* for nitrogen fixation following root hairs penetration.

The likeliest explanation for the decline of *rhizobial* population is grazing by protozoa. The data show that these organisms replicate rapidly as the seeds germinate, and they reach numbers sufficiently large to account for decline in counts of total bacteria as well as *rhizobia.* Benomyl application on seed drastically reduces the decline of *rhizobia* around soybean roots is because of inhibition of protozoa by benomyl. Similar to this thiram also inhibit protozoa and increased the *rhizobial* counts around bean roots. Protozoa have an anologus influence when added to sterile soil together with the mixture of other micro-organisms.

Broad spectrum fungicides applied to soil or seed not only have an effect on root nodule bacteria, but also they may influence the interaction among the soil micro-organisms and *rhizobia.* An inhibition by the fungicides of soil or *rhizosphere* inhabitants that complete with lyse, produce inhibitors against, parasitize or prey on *rhizobia* may prove beneficial because suppression might overcome the natural biological control of the *rhizobia.* Hence, a fungicide resistant strain of *Rhizobium* might proliferate more extensively in the legume *rhizosphere* and nodulate more frequently should its potential antagonists be inhibited by the added fungicides.

It is concluded from the several studies that grazing by protozoa and competition with bacteria limit the growth of *rhizobia* in soil apart from plants. Regardless of the precise contribution of proliferation on root surfaces and in soil under the direct influence of plants, the results of this study show that colonization can be enhanced by suppressing the predators or competitors by use of antimicrobial agents. Benlate and Bavistin systematic fungicides-Benlate or Bavistin at 0.15 per cent and non-systemic fungicides-thiram, captan and mancozeb at 0.3 per cent are being recommended for the control of seed borne diseases. Besides these antibiotics are also used for the seed treatment for the control of bacterial diseases. It is recommended that in seed treatment of legumes judicious use of these fungicide should be made.

Stem Modules

Certain aquatic and water tolerant legumes are known to bear nitrogen fixing nodules on stems. These stem nodules are caused by *Rhizobium.* Only some species of *Aeschnomene* and' *Sesbania* are known to bear stem nodules and have been investigated in this regard. Recently, renewed interest on these stem nodulating plants has been evidenced in view of the potential benefit that could be achieved, if similar stem nodulation of legumes, N-fixing stem nodules have been recorded in warry lecellate barks of various trees inhabited by entero-bacteria. Stem nodules offer a challenging and convenient system to investigate the inter-relationship between photosynthesis and BNF. The cortex of stem nodules is green due to chlorophyll and the bacteriod tissue is pink possibly due to leghaemoglobin tissue for the process of nitrogen fixation can easily be traced and measured with accuracy by the use of labelled carbon. The stem nodules can easily be peeled off at the base for laboratory experiments. The genetic basis for the stem nodulation habit may have to be understood in great detail so as to induce stem nodules in cultivated crops to secure better BNF of atmosphere N.

The bacterium, *Aeschynomene indica,* a gram-negative, coccoid prokarryote wan elaborate internal membrane system reminiscent of certain photosynthetic bacteria has been discovered. Ultra structurally, this organism bears a striking resemblance to the purple photosynthetic bacteria of the genus *Rhodopseudomonas,* such as *R. capsuoata* and *R. spharoides.*

Genus-Frankia

The genus *Frankia* belongs to family Frankiaceae order Actinomycetales. The associations between non-legumes and *Frankia* have been given the name actinorhizal associations. Because this association is far less specific than *Rhizobium,* current practice is to refer to the generic name only *(i.e. Frankia)* without attempting to define the species. Dr. Callatiam and Cownkers from UK cultivated *Frankia* for the first time in 1978 in pure culture. The actinomycete could reduce acetylene was a further advance, when grown in complex liquid media containing fixed N compound. *Frankia* proliferates as a filamentons mat consisting of branched, septate hyphae. However, in response to N deprivation, short, branched filaments develop into swollen terminal vesicles which are capable of reducing acetylene. These spherical vesicles, about 3-5 mm in diameter are developed by *Frankia* which have so far been examined in N-limiting pure culture. However, within the host, a variety of vesicle shape is detected, varying from the spherical vesicles observed in pure culture to complete absence of vesicles. Studies on the morphology of nodules in various host indicate different shape and separations. The uniform appearance of vesicles in pure culture, therefore, contrast with the variation in vesicle morphology which occurs in the host and suggests that the host may have a part to play in determining vesicle shape. The vesicles possess an outer layer called the vesicle envelop, which is multilaminate. The envelop functions as a barrier to the diffusion of O_2 into the vesicle and thus protects nitrogenase from inactivationby O_2. The concentrations of oxygen that determines vesicle occurrences and shape.

A number of non-legurninous shrubs and trees (e.g. *Alnus, Myrics, Hippophae, Casurina* and *Eleagnus*) are nodulated by the nitrogen fixing Actinomycetes, *Frankia,* with Frankiaceae (Actinomycetales), the only actinomycete reported to fix nitrogen to date. The so-called actinorhizal plants include 173 species of seventeen genera belonging to eight families and the number is growing all the time so the range of hosts is much more wider than for *Rhizobium.* Although there is some host specificity *Frankia* is much more promiscuous in the range of hosts infected than in Rhizobium. For this reason no species have yet been created.

Frankia nitrogen fixing activity has been shown in a free living stage *in vitro* but its actual contribution to soil nitrogen economy comes mainly through its capacity to establishing nitrogen fixing symbiosis with over 200 plant species all of them of a woody nature but for two species of *Datisca,* non-leguminous higher plants belonging to some 23 genera. The nitrogen fixed in the root nodules of these symbiosis is in part translocated eventually to the rest of the plant. Then the nitrogen goes back to the soils in the annual decomposition of the plant material and through the root exudates, both important for the growth of associated plant communities. *Frankia siimbiosis* also provides other nutrients to the plants, which contribute to soil fertility. Deciduous trees conserve nitrogen by mobilizing leaf nitrogen in autumn with subsequent storage of the nutrient in twigs and branches, from where it is made available for growth the following spring. Such a phenomenon is not observed in alders plants probably owing to the fact that they can fix atmospheric nitrogen and obtain lesser benefits from such conservation mechanisms. Experimental evidence indicates that much nitrogen is present in the leaves at leaf fall.

Actinornizae are pioneer species of poor soils, and at present their commercial potential is restricted to forestry and land reclamation. Some of the trees, such as alder are used in agriculture as wind breaks. Rates of nitrogen fixed can be comparable to legumes (upto 150 kg nitrogen per hectare per year for alder) but they are frequently less.

Nodulation seems to follow a pattern similar to that observed in legumes but is much less clearly understood. Nitrogen fixation occurs in terminal swellings of the actinomycete hyphae, called vesicles. Each infected cell contains one vesicle cluster containing several hundred spherical vesicles. Nodule size can range from few millimetres to several centimeters in diameter. They are not pink in colour, but a leghaemoglobin-like molecule has recently been detected in nodules from most but not, all actinorhizae investigated. It is recently proved possible to grow *Frankia ex planta* in pure culture. Although it is often described as a microaerophile, nitrogen fixation in pure culture appears to be relatively insensitive to oxygen. All the isolations have been made from nodules. To date no-one has succeeded in isolating *Frankia* from soil stage in its life cycle. All isolates produce stores in large sporangia (upto 60 mm in diameter) when grown in pure culture, but only some of them produce spores inside the nodule.

Biofertilizer Role

A biofertilizer role of actinorhizal plants has been demonstrated by the works of Dr. G. Bond Glasgo University) and Dr. A.I. Virtanen of

Helsinki University, using isotopic N_2 determination. They reported that a sizable amount of N_2 was released from the biomass of actinorhizal plants from ihe amount of nutrients incorporated to the soil through leaf fall. Leaf nitrogen from actinorhizal r lants ranges from 1.8 per cent to 3 per cent. The N_2 content at leaf fall lowers C/N ratio in litter leading to its faster decomposition in soil and making the nutrients, available to other crops growth.

Alder leaf litter takes one to four years to decompose completely whereas it takes some seven to ten years for a complete decomposition of pine needles. In Myrica gale decomposes 40 per cent biomass and releases 10 per cent N_2 after five years. The slow rate of decomposition is due to higher legnin content of this plants. The *Fmnkia* nodulated plants possess a nitrogen reductase system which absorbs mineral nitrogen and stores in the leaves, roots and nodules.

Genus-Azolla

Introduction

Azolla is a small aquatic fern generally found is omnipresent in nature in shallow ponds with stagnant water in winter season. Morphology of *azolla* consists of a floating branched rhizome with small alternate bitobed leaves and roots hanging down in water. Although spores are present but reproduction is mostly through vegetation. The important characteristics of *azolla* is that a blue green algae (BGA), each leaf has dorsal and ventral lobe. The dorsal lobe is green and has a blue green algal symbiont *(Anabaena azollae)* within a central cavity. The heterocyst of the symbiont *Anabaeria* is the site of N_2 fixation.

The congenial temperature for the growth of *azolla* lies between 15°C to 30°C. Favourable pH is 6 to 7. It grows less at alkaline pH and stops growing in acidic soil. It can be grown in shallow ponds or low lying lands during October to February in plain. A layer of *azolla* growth over one hectare of land produces about 10 tonnes of green matter which may give about 30 kg N/ha for growing rice. *Azolla* is very rich in nitrogen containing about 4 to 6 per cent N on dry weight basis. It also has sufficient quantity of phosphorus and potassium. Average chemical composition of *azolla* on dry weight basis is given in Table 4.4.

Table 4.4: Average chemical composition of *Azolla*

Nutrient	On Dry Weight Basis Per cent Content
Nitrogen	4.5
Calcium	0.7
Iron	0.16
Manganese	0.15
Magnesium	0.55

The importance of *azolla* as an organic input in rice cultivation was first demonstratrated in North Vietnam in 1957 and subsequently its potentiality has been recognized in U.S.A., Indonesia, Japan, Philippines and China respectively. *Azolla* grown in the rice fields and ploughed under, so as to facilitate its quick decomposition, however, in some areas it is grown as a dual crop along with the rice. *Azolla* plants are triangular or polygonal in shape and the leaves consists of green or purple chlorophyllous dorsal lobe which contains algal symbionts (BGA). *Azolla pinnata* is commonly found in India, out of six sp. distributed throughout the world and it has been seen that this sp. from Vietnam grows well in summer as it keeps lustre green vigour throughout the year. Under most favourable conditions, it has been observed that as a result of symbiosis, it has been observed that as a result of symbiosis between blue green algae and *Azolla* upto 900 kg N/ha/year is fixed. Phosphorus application happens to be one of the most critical factor to exploit its full potential. Different strains have differential nitrogen fixing ability depending upon agro-climatic conditions.

Morphology and Taxonomy

The genus *Azolla* belongs to the order Salviniales has been assigned to a monotypic family Azollaceae and subdivided into two genera on the basis of their reproductive morphology *viz., Eiiazolla* and *Rhizosperma,* There is no morphological differences in the species and they are widely distributed in tropical and temperate fresh water ecosystem throughout the world. *Azolla* is indigenous to some rice fields and can grown in others, when inoculated. The various species *viz., A. caroliniaba* (North America), *A. filiculoides* (South America), *A. microphylla* (tropical and subtropical America) are belongs to subgenus *Euazolla; A. pinnats* is distributed in tropical and subtropical Asia and coastal Africa and *A. nilotica* (Africa) are belongs to subgenus *Rhizosperma.* The individual *Azolla* plants are 1-5 cm but *A. nilotica* which becomes as long as 15 cm. The plant consists of a branched floating stem that bears alternatively arranged overlapping leaves and true roots. Each leaf has two lobes, the dorsal and ventral. The later lobe contains and external cavity occupied by the Anabaena azollae symbiont. This is the species known to be symbiotically associated with Azolla and belongs to phylum cyanophyta, order Nostocales and non-random distribution suggest the possibility of recognition site in the Azolla. Anabaena azollae increases from apex to base in the successive leaves concomitantly the nitrogen fixation capacity increases. Alga free Azolla has rarely been found in nature.

Azolla plants are triangular and polygonal in shape consists of floating branched rhizomes bearing small alternative overlapping leaves and simple roots. The bilobed leaves consist of green or purple chlorophyllous dorsal lobe which contains the algal symbiont (BGA) within the internal cavity and achlorophyllous boat shaped ventral lobes which contain spongy tissues, symbiont BGA and *Azolla.* The symbiotic BGA occurs as unbranched filamentous consists of vegetative cells and heterocysts. The heterocysts possess a complex wall sheltered and multi-enzyme system called nitrogenase fixes atmospheric nitrogen and supply to the *Azolla* and in return gets the photosynthetic products from *Azolla.*

Role of *Azolla*

Azolla has a unique role as biofertiliser:

1. In replacing/supplementing chemical fertilisers to meet nitrogen for increasing rice yield.
2. As organic input in rice cultivation.
3. It has the potential to create self-employment among educated rural youth through its scientific production and distribution system in India.

Due to continued research in India and abroad the Blue Green Algae (BGA) and *Azolla* biofertiliser has been identified for rice cultivation.

It has been reported that single incorporation of *Azolla* in rice field is comparable to 30 kg N/ha. *Azolla* incorporation significantly increases total soil nitrogen, organic carbon, available phosphorus. Moreover, it improves the soil physical condition, *i.e.,* decreases specific gravity and increases porosity. Its weed suppressing potential and beneficial residual effects to succeeding crops make it more popular. Under favourable environmental condition like stagnant water, temperature (20-30°C), light energy at visible spectrum, all macro and micro nutrients except nitrogen, *Azolla* found to grow better. It can tolerate the salt concentration of 90 to 15 mg salt/lit.

Inoculum Production of *Azolla*

In order to exploit the multiple benefits of *azolla,* particularly as a biofertiliser in rice production the primary need is to produce large quantity of inoculum in the vicinity of the rice fields with least application of inputs and without involving productive crop land. A simple and low-cost technique - as been developed to produce large quantity of both vegetative and spore inocula. As a result, 2.5-5.5 tonnes/ha/day biomass can be produced in hypereutrophic water bodies under the partial canopy of suitable trees during October through March. Simultaneously, biomass-based spore inoculum can be produced in open area of the water bodies during January through March. The spore production becomes of maximum, when the plant turns brownish in open area. The spores germinate after 3 months of storage and remain viable for more than 12 months, however, germinability is reduced by 20-30 per cent after 9 months of storage. Several institutions has started producing vegetative inocula for the local farmers and spore inocula for the distant farmers. Though the vegetative inoculum has the ability to produce larger biomass within a short time, spore has also disadvantages. It can be used directly in the rice field or as a means of producing vegetative inoculum. It can tide over the unfavorable seasons of winter and summer and it could be transported over a long distance. The cost of production of vegetative inoculum is Rs. 25 per 250 kg packet, that of one kg biomass-based spore inoculum packet is Rs. 1 while that of spore only is Rs. 25 per 100 gm packet.

There are plenty hyper-eutropic water bodies in different parts of India, particularly in eastern India. So, rural people, particularly women can utilise both the biomass production processes either for vegetative inoculum or spore inoculum to develop village industries at a very low cost without involving additional land, water, fertilisers, etc. but recycling a considerable amount of nutrient from the aquatic system to the crop land wherefrom the nutrients are mostly lost. However, one has to take care while utilising the water bodies.

The oxygen concentration and phytoplankton density are not reduced with half coverage of the *azolla* mat. So, the *azolla* biomass should be harvested when the mat develops half coverage to avoid any deleterious consequences on the aquatic animals. Even the farmers can produce small quantity inoculum in the water bodies by fencing a small part of the water bodies through a floating chord.

Factors Affecting Successful *Azolla* Production

The most important factor limiting the growth of *Azolla* is temperature as according to FAO report the optimum temperature for proper growth of *Azolla* stands in between 20 to 30°C. Out of six sp. of *Azolla-A. carotiniana, A. nilotica, A.filicldoides, A. mexicana, A. microphylla* and *A. pinnata,* last sp. has been found most tolerant sp. to high temperature and competent under Indian conditions. As pH of the soil is another limiting factor as pH range of 6-7 is most suitable but according to work done at the CRRI, Cuttack, ideal pH for this purpose should be around 8.0. Standing water of 5-10 cm deep and application of super-phosphate at the rate of 5-10 kg/ha seemed essential for rapid growth of *Azolla.* To avoid *Azolla* nurseries from wind erosion, it is preferable to use plots of size 50 ´ 100 sq.m. than bigger plots. Use of carbofuran at the rate of 1-2 kg/ha is equally important to prevent the spread of insect parasites and the consequent destruction of *Azolla* is further strengthened either because of poor quality of water or because of unavailability of water.

Azolla Nursery

A simple rural oriented low cost method for raising *Azolla* inoculum for rice has been recommended. The details of multiplication from the starter culture include:

1. Inoculating 1.0 kg of fresh *Azolla* in 10 sq.m. plot having 10-30 cm standing water, apply 30 gms of single super phosphate and 1 gm a.i. carbofuran.
2. On 5th day apply 30 gm single super phosphate as foliar spray (1 per cent solution).
3. On 10^{lh} day harvest *Azolla* (10-15 kg) and inoculate in 100^{2} in size plot having 10-30 cm standing water, apply 300 gms of single super phosphate and 10 gms a.i. carbofuran.
4. On 15th day apply 300 gms of single super phosphate as foliar spray (1 per cent solution).
5. On 20th day harvest 100-400 kg of *Azolla* and inoculate it to a field of 1000 m^{2} size (0.1 ha) having 10 cm of standing water. Apply 3 kgs of single super phosphate and 100 gms a.i. carbofuran.

6. On 25th day apply 3 kg of single super phosphate as foliar spray on *Azolla.*
7. On 30th day 1 to 1.5 tonnes fresh *Azolla* is ready to harvest, which is sufficient to 1 ha of rice field.

Recently a simple technique has been developed to produce dried sporocarp inoculum for *Azolla* which has a shelf life of two years. Pre-treatment of dried sporocarp with growth regulators increases germination and enhances the growth of young sporophytes. Only 200 kg fresh weight of *Azolla* fronds is sufficient per hectare.

Field studies carried out under the auspices of All India Coordinated Agronomic Research Project (AICARP) of ICAR at different locations, have revealed that incorporation of 5 t/ha of *Azolla* gave *Azolla* yield equivalent to about 30 kg N/h at Bhubneshwar, Khaiagpur.

Azolla Nursery method has been evolved for its large-scale multiplication in the field forestry adoption by the cultivators. The field selected for its nursery must be thoroughly prepared and levelled uniformly and then divided into 20 m ´ 2 m plots by providing suitable bunds and irrigation channels. Water is maintained at a depth of 10 cm. Ten kg fresh cow dung mixed in 20 litres of water is sprinkled and *Azolla* inoculum at 8 kg is dispensed to each plot. Single super phosphate (100 g) is applied in three split doses at 4-day interval as a top dressing for *Azolla.* For insect pest control, corbofuran, granules at 100 g/plot is applied 7 days after inoculating field as source of primary inoculum. From one harvest, 40-55 kg fresh *Azolla* is obtained from each plot. Again, it is inoculated in the same plot and cow dung single superphosphate and carbofuran are applied. Four cents nursery plots are required for raising *Azolla* for inoculation to one hectare main field. Adding cowdung in flooded rice soil system is reported to increase its growth and nitrogenase activity. The use of cowdung and cattle-shed water increases its growth during winter.

Constraints

The biggest constraint in the large scale use of *Azolla* is inadequate availability of its location specific strains. Also, its multiplication is a encumbersome process as it has to be grown and multiplied in separate ponds as an independent crop, requiring lot of assured supply of water and its growth is much influenced by temperature fluctuations. In fact, it has to be raised as a crop, requiring fertilisation and proper care. Because of these limitations, *Azolla* seems to have limited potential and raising of a crop of *Azolla* often turns out to be a specialised job, especially in regions with high temperature. The growth and efficiency of *Azolla* in general is high in hilly or semi-hilly humid areas or in coastal areas than in regions with high temperature.

Conclusions and Future Outlook

Azolla biofertiliser has a tremendous significance on the future scenario of fertilizer market. It is one of the best alternative source of Nitrogen and organic matter. Moreover, this technology may be manipulated to develop "Dry *Azolla* Technology" and to evolve stains which can sustain high temperature and low phosphorus. On weight basis, the yearly *Azolla* requirement of our country is approximately 400 million tonnes. This formidable figure needs a massive extension efforts, training and participation of farmers and entrepreneurs, adequate investment from Government and Non-Government agencies to achieve this target.

China and Vietnam are the leading countries in world where *azolla* application in rice cultivation has been a common practice since-long. Ze Jiang University at Hong Zhou (China) is the leading centre of *azolla* research. Realizing the role of *azolla* in Agriculture University of Hawaii has recently initiated an International *Azolla* Project being coordinated by Dr. Tom Lumpkin, a leading International *Azolla* Research Scientist. The positive effect of *azolla*, an experiment on use of *azolla* as an organ nitrogenous fertiliser for rice was chalked out in All India Coordinated Research Projects and results revealed that incorporation of *azolla* @ 5 to 10 t/ha was equivalent to the application of 300 N/ha as inorganic nitrogen fertiliser.

However, there is a great need to conduct trials on large scale both at research centres and farmers fields by the Agricultural Universities / Institutes of the rice growing areas to explore possibilities of practical use of *azolla* in rice cultivation in the country.

Though *Azolla* is known to be a very efficient green manure for rice, its use is restricted areas, where water is available in off-seasons. Further, high temperature and severe winter in some regions prohibit its rapid multiplication and adoption technique. Identification of tolerant strains development of cheap and efficient techniques for their multiplication are important requirements promoting the technology. In addition, an effective network for *Azolla* production and distribution has to be initiated to facilitate its large scale adoption, in rice-growing regions.

Integrated Approach for Increasing Microbial Inputs

In bio-farming biotic agents play very important role in integrated soil fertility manganese (ISFM). These soil biota (biotic agents) have the ability to carry out biochemical transformations as serve as a source and sink for mineral nutrients. A range soil biota are involved in various beneficial traiisf ormation processes, such as BNF, mineralization of nutrients, organic matter decomposition increased nutrient uptake and use efficiency, etc. This is being optioned by several microbiologist that the micro-organisms are not being exploited efficiently in agriculture for increasing production and productivity to a expected level.

Micro-organisms in agriculture use and their activities governed by several facts, such as micro-organisms per se, their mechanisms of benefits, host controlled factors and lastly appropriate management practices and each factor is depended on other, hence it is suggested all have to integrate achieve maximum benefits. In this connection it is appropriate to use *Rhizobium/Bradyrhizobhim* an example to study the ways for increasing the benefits in biof arming. Several attempts have been made to select the most efficient strains of the biofertilising agents. Biological nitrogen fixation (BNF) for example involve host, bacterium and environment. When more than one organisms are involved, it becomes necessary to study the system involving major groups of beneficial organisms, so synergistically deliver maximum benefits.

The fundamental knowledge on the mechanisms by which the introduced micro-organisms benefit the crop is very critical. Besides, the major activities of the biofertilising agents, studies have shown that other functions, such as production of siderophores, hormones or antibiotics or increases productivity through increased root growth also help plants to increase yield. Further, there is need to test strains of one group of organisms, such as nitrogen fixers, VAM, PSM, etc. which can work in synergistic way in the soil. It is of prime importance that the activities of micro-organisms is to be improve in field conditions. The survival and competitive ability of the strains to be introduced by improved. There is also need for identifying biological, chemical and physical factors that influenced establishment of the inoculated agents in the rhizosphere and full expression of its trait for benefiting the crop. The host factors, such as genes, physiology, specificity and selectivity are also important.

Further there is need to identify type of legumes and also genotypes a given legume which yield more and also derive larger benefit cost the proportion of their nitrogen requirement through BNF in a particular cropping system. As low nodulating and non-nodulating plants were found in the existing cultivars and/or segregating populations, efforts are required to minimise the occurrence of such plants. Suitable management practices need to be provided to ensure maximum contribution from the inoculants. It is generally observed that VAM provide maximum benefit in low phosphorus situations, nitrogen fixers perform better under low soil nitrogen situation, etc. Nitrogen application should be manipulated in such a way that in intercropped situations slow released nitrogenous fertilisers should be used and these to be placed at the root zone of cereal crops rather than broadcasted, Application of organic matter improve the efficiency of the inoculants, Stagnation of water should be avoided and also water conservation practices should also be followed.

Economics of Biofertiliser Use

Inconsistency in crop response to biofertilisers is a rule rather than an exception. However, the cost of biofertilisers is so low that even small increases in crop yields, though not significant, will be in excess of likely cost of inoculation which would make this technology attractive to the farmers. Monetary gains from the use of biofertiliser was reported. The investment of one Rupee in *Azolla* dual cropping gave a net return of Rs. 21-24 as against a net return of Rs. 5 with fertiliser N. A net income of 2,000/- from 0.2 ha land by using BGA biofertiliser technology has been reported. Based on the field trials by ICAR and SAU in different agro-climates estimation of the biofertiliser equivalent of important biofertilisers has been made. According to that, fertiliser equivalent of 19-22 kg N/ha from *Rhizobium,* 20 kg N/ha from *Azotobacter* and *Azospirillum,* 20-30 kg N/ha from BGA and 3-4 kg N/ha of *Azolla* could be realised. If one takes into consideration the cost of fertiliser N equivalents of biofertilisers and the cost of fertiliser, it will certainly be very economical to use biofertiliser.

Obviously, even if half of the indicated potential of each biofertiliser could be realised under normal field conditions, their use would be highly cost effective.

Besides their direct effect on current crop, use of a biofertiliser also leaves considerable beneficial effect on soil fertility observed that yield of various Rabi legume fodders increased by 10-35 per cent with the aid of *Rhizobium* inoculation and the yield of subsequent Kharif sorghum fodder raised on residual fertility was also 17-33 per cent higher than the control; non-inoculated plots. The residual effect of a leguminous crop inoculated with a biofertiliser on the succeeding cereal crop is believed to range between 20 to 123 kg N/ha.

Rhizobium possesses better tolerance to salt and pH (5 to 10) under various ecological conditions, therefore, possesses better adaptability to different agro-climatic situations. Biofertilizers may exert favorable effect on root growth and crop stand by affecting general growth and development of plant and sustained greenness. They may hasten flowering and crop maturity to a certain extent. Soil application of 2.5 kg/ha of *Azospirillum* hastened flowering and maturity of sunflower by 15 days. The crop was also observed to withstand favorably the impact of dry spell towards later stage of the 1 crop. *Azospirillum* and Phosphobacterin produce growth promoting substances which have been reported to promote number and length of roots, besides shoot growth.

Use of biofertilisers is economical with a high cost, benefit ratio, without risk. Soil application of 800 gm of *Azospirillum* was found to give, on an average, yield increase of rice by 1500 kg/ha resulting in net profit of 3250/- per hectare.

If the strain of bio-inoculant is properly evolved, it might work simultaneously as a "biofertiliser" and a "biopesticide", as was observed in the case of *Azotobacter* strain 1006 by Shah and Joshi. The said strain was claimed to have shown potential in inhibiting seed borne pathogens in some cereals.

Biofertilisers are renewable sources of plant nutrients and supplements fertiliser requirement in crop production. The available data indicate that biofertiliser can supply upto 20 per cent of the nitrogen requirement of crops, Research findings have established that upto 2 kg of biofertilisers applied to one ha will fix 25 kg of atmospheric nitrogen in the soil. In term of cost per acre, biofertilisers worth Rs. 20/- applied to the crop reduces the requirement on urea worth Rs. 175/- even at the current prices. Hence, there is a very good economic incentive for the farmer to supplement chemical fertilisers with biofertiliser.

Outlook

This is conclusively proved that *Rhizobium* inoculation of leguminous plants is highly effective in exploiting and harnessing atmospheric nitrogen for improving the nitrogen status of the soils. Symbiotic nitrogen fixation could occur only within the cross inoculation group. Nitrogen fixation by biological agents as distinguished from chemical N fixation in a fertilizer factory. BNF is mediated by several organisms. The ratio between chemically fixed N and BNF ranges approximately from 1:4 to 1:2:5. But Rhizobium fixes half of the total BNF to that of industrial fixation. However, the symbiotic N_2-fixation is adversely affected by adverse soil conditions. Seed pelleting is effective in protecting the inoculated seeds against adverse conditions like acidity, sodicity and pesticides. The quantity of nitrogen varies from 40-200 Kg/ha/year. The nitrogen fixed symbiotically not only benefits the host but also has a residual effect to benefit the subsequent crops.

□□□

5

Farming System and Sustainable Agriculture

Progress of any country depends upon the prosperity of farmers. This in turn depends upon the adoption of improved technology and judicious allocation of resources. Human race depends more on farm products for their existence than anything else since food and clothing—the prime necessaries are products of farming. Even for industrial prosperity, farming forms the basic raw material. Majority of the farmers are cultivating their lands without the backup of scientific skills. To make the farming economically viable and environmentally sound and sustainable, a holistic all around approach is a must. To meet the multiple objectives of poverty reduction, food and nutrition security, competitiveness and sustainability, several researchers have recommended the farming system approach.

A farming system is the result of complex interactions among a number of inter-dependent enterprises/components, where an individual farmer allocates certain quantities and qualities of four factors of production, namely land, labour, capital and management to which he has access. Farming system approach is a powerful tool for natural and human resource management in developing countries, such as India. It is a multi-disciplinary whole- farm approach and can be effectively employed in solving the problems of small and marginal farmers. The approach aims at increasing employment and income from small holdings by integrating various farm enterprises and recycling crop residues and by-products within the farm itself. In view of the decline in per capita availability of land from 0.5 ha in 1950-51 to 0.15 ha by the turn of the century and a projected further decline to less than 0.1 ha by 2020, it is imperative to develop strategies and agricultural technologies that enable adequate employment and income generation, especially for small and marginal farmers who constitutes more than 80 per cent of the farming community.

Under the gradual shrinking of land holding, it is necessary to integrate land based enterprises like fishery, poultry, duckery, apiary, field and horticultural crops, etc. within the bio-physical and socio-economic environment of the farmers to make farming more profitable and dependable. No single farm enterprise is likely to be able to sustain the small and marginal farmers without resorting to integrated farming system (IFS) for the generation of adequate income and gainful employment year round. Farming system approach, therefore, is a valuable approach to address problems of sustainable economic growth for fanning communities in India. The basic aim of IFS is to derive a set of resource development and utilisation practices, which lead to substantial and sustained increase in agricultural production.

Farming is a process of harnessing solar energy in the form of economic plant and animal products. A **system** is a set of inter-related, interacting and interdependent elements acting together for a common purpose and capable of reacting as a whole to external stimuli. It is unaffected by its own output and it has external boundaries based on all significant feed backs. Farms are system because several activities are closely related to each other by the common use of the farm labour, land and capital, by risk distribution and by the joint use of the farmer's management capacity.

Farming system for sustainable crop production intensification will offer a range of productivity, socio-economic and environmental benefits to producers and to society at large, including high and stable production and profitability; adaptation and reduced vulnerability to **climate change;** enhanced **ecosystem** functioning and services and reductions in agriculture's **greenhouse gas** emissions and **"carbon footprint".** These farming systems will be based on three technical principles :

1. Simultaneous achievement of increased agricultural productivity and enhancement of natural capital and ecosystem services.
2. Higher rates of efficiency in the use of key inputs, including water, nutrients, pesticides, energy, land and labour.
3. Use of managed and natural **bio-diversity** to build system resilience to abiotic, biotic and economic stresses.

The farming practices required to implement those principles will differ according to local conditions and needs. However, in all cases they will need to :

- Minimise soil disturbance by minimising mechanical tillage *in order to maintain soil organic matter, soil structure & overall* ***soil's health.***
- Enhance and maintain a protective organic cover on the soil surface, *using crops,* ***cover crops*** *or* ***crop residues,*** *in order to protect the soil surface, conserve water and nutrients, promote soil biological activity and contribute to integrated weed and pest management.*
- Cultivate a wider range of plant species - *both annuals and perennials- in associations, sequences and rotations that can include trees, shrubs, pastures and crops, in order to enhance crop nutrition and improve system resilience.*

1.1 Definitions

Farming system approach relates to the whole farm rather than individual elements; it is driven as much by the overall welfare of farming households as by goals of yield and profitability.

Farming system refers to an ordered combination of crops grown, livestock produced, husbandry methods and cultural practices followed. A farming system is defined as a population of individual farm system that have broadly similar resource bases, enterprise patterns, household livelihoods and constraints and for which similar development strategies and interventions would be appropriate (FAO–Farming system).

Farming system is an integrated set of activities that farmers perform in their farms under their resources and circumstances to maximise the productivity and net farm income on a sustainable basis. Farming system takes into account the components of soil, water, crops, livestock, labour, capital, energy and other resources, with the farm family at the centre managing- agriculture and related activities.

Farming system is defined as a complex inter-related matrix of soil, plants, animals, implements, power, labour capital and other inputs controlled in part bv farming families and influenced to varying degrees by political, economic, institutional and social forces that operate at many levels. Fanning system, therefore, refers to the farm as an entity of inter-dependent farming enterprises carried out on the farm. The farm is viewed in a holistic manner. The farmers are subjected to many socio-economic; bio-physical, institutional, administrative and technological constraints. Other definitions of farming system include:

Farming system is a resource management strategy to achieve economic and sustained agricultural production to meet diverse requirements of farm livelihood while preserving resource base and maintaining a high level of **environment quality** (Lai and Miller, 1990).

Farming system is a set of agro-economic activities that are inter-related and interact with themselves in a particular agrarian setting. It is a mix of farm enterprises to which farm families allocate its resources in order to efficiently utilise the existing enterprises for increasing the productivity and profitability of the farm. These farm enterprises are crop, livestock, aquaculture, agroforestry and agri-horticulture (Sharma *et al.*, 1991).

Farming system is a mix of farm enterprises, such as crop, livestock, aquaculture, agro-forestry and fruit crops to which farm family allocates its resources in order to efficiently manage the existing environment for the attainment of the family goal (Pandey *et al.*, 1992).

Farming system represents an appropriate combination of farm enterprises (cropping system horticulture, livestock, fishery, forestry and poultry) and the means available to the farmer to raise them for profitability. It interacts adequately with environment without dislocating the ecological and socio-economic balance on one hand and attempts to meet the national goals on the other (Jayanthi *et al.*, 2002).

Farming system is a decision making unit comprising the farm household, cropping and livestock system that transform land, capital and labour into useful products that can be consumed or sold (Fresco and Westphal, 1988).

1.2 Scope of Farming System

Farming enterprises include crop, livestock, poultry, fish, sericulture etc. A combination of one or more enterprises with cropping when carefully chosen, planned and executed gives greater dividends than a single enterprise, especially for small and marginal farmers. Farm as a unit is to be considered and planned for effective integration of the enterprises to be combined with crop production activity.

The need for **farming system approach** in the present scenario is mainly due to high cost of farm inputs, fluctuation in the market price of farm produce, risk in crop harvest due to climatic vagaries and biotic factors. **Environmental degradation,** depletion in **soil fertility** and productivity, unstable income of the farmer, fragmentation of holdings and low standard of living add to the intensity of the problem.

Farming system approach is an approach for developing farm-household system, built on the principles of productivity, profitability, stability and sustainability. All the components are **complimentary** and **supplementary** to each other. The development process involves participation ol rural communities. Fanning system approach emphasizes understanding of farm household, community inter-linkages, reviews constraints and assesses potentials. It combines improvements desired from better technology. It needs efficient support services and requires better policies. It is continuous, dynamic and interactive learning process based on analysis, planning, testing, monitoring and evaluation. Fanning system approach is necessary :

- To develop farm-house hold system and rural communities on a sustainable basis.
- To improve efficiency in farm production.
- To raise farm and family income.
- To increase welfare of farm families and satisfy basic needs.

Farming system addresses two issues: reduction in risk with **monoculture** activities and promoting enterprise diversification, value addition and development of alternate income sources with efficient utilisation of farm resources. It brings about enterprise diversification for **sustainability** and additional benefits, better management of important farm resources like land, labour, capital, etc. The FS provides an opportunity for effective recycling of the products and by-products. It helps to generate flow of cash to the farmers round the year by way of disposal of milk, fruits, fuel, manure, etc., beside other agricultural output.

1.3 Farming System Strategy

In view of serious limitations on horizontal expansion of land and agriculture, only alternative left is vertical expansion through various farm enterprises requiring less space and time but giving high productivity and ensuring periodic income, specially, for the small and marginal farmers located in **rainfed areas,** drylands, arid zone, hilly areas, tribal belts and **problem soils.** The following farm enterprises could be combined:

- Agriculture alone with different crop combinations.
- Agriculture + livestock.
- Agriculture + livestock + poultry.

- Agriculture + horticulture + sericulture.
- Agro-forestry + silvipastoral.
- Agriculture (rice) + fish culture.
- Agriculture (rice) + fish + mushroom cultivation.
- Floriculture + apiary (bee-keeping).
- Fishery + duckery + poultry, etc.

For meaningful execution of integrated farm enterprises, the following activities should be undertaken by a multi-disciplinary team of extension professionals with farmer's participation and involvement at all stages :

1. Thorough understanding of existing farming systems and their components.
2. Assessment of resource availability in the farm environment and identification of bio-physical, socio-economic, institutional, administrative and technological constraints.
3. Developments of economic viable and efficient **integrated farming system** suitable for various domains.
4. Diffusion of improved technology and receiving 'feed back' for further improvement of the system as a whole.
5. Continuous improvement in components technology to fit into a given farming system.
6. Improvement in quality of farming system.
7. Research-extension linkage through on-farm adaptive research.
8. Development of national and international linkages.

1.4 Concept of Farming System

The newly developed concept of integrated farming is a synthesis of our age-old traditional knowledge and modem scientific techniques.

Early in our civilization, the source of food was mainly through collection or hunting. As people started to settle down as communities in villages, they felt the need for a more stable food. The inspiration for our ancestors was natural production system - pastures and forests. They emulated the natural production system as they started growing food instead of hunting and gathering. Such agricultural system had the diversity alike any natural system. Such a set up was self-contained and self-sufficient i.e. it did not require any input from outside the farm or village. Such systems used to give enough to feed the family. Slowly, through the ages, our farmers developed a rich body of knowledge about how to produce while in tune with nature and without harming the nature.

But our traditional technologies did not produce enough. Way back to sixties, our country was going through food crisis. As we could no longer expand our arable lands much, the challenge was to increase the productivity of our farmlands to feed the fast growing population. Traditional agriculture failed to meet the food needs and there was a necessity for new technologies. Investigations revealed that the low productivity was due to low external input traditional agriculture.

A high external input based agricultural system was pushed ahead to salve the problem. The era of **"green revolution"** began. Green revolution hiked the productivity of our farmlsnds many fold. Countrie's food stores swelled over the next few decades. But what happened to the green revolution? There are again serious crices in our food production today. A whole range of problems suddenly plunge our agriculture. Productivities are falling. Farmers are commiting suicides. The countrie's population continues to grow as before. There are serious concerns all over again today whether we shall be able to feed our teeming millians in a decade from now?

What has gone wrong? Rampant use of chemical fertilisers had led to the steady decline of **organic matter** in soil and the soil has lost its "life". Our natural ecosystems are poisoned -- the food that we eat is poisoned and every day there are newer pests and pothogens in our crop fields that have grown resistance to pests and pathogens. Our groundwater store is declining fast. Green revolution prompted our farmers to trap the aquifar for crop production. Our rich crop diversity is lost due to few high yielding cultivars that now sway in oue farmlands.

The small and marginal farmers today are at their wits end. Market dictate them today about what to grow and when not to eat, what they grow but sell in the market to get some cash in hand to by food from the market.

Solution to whole range of problems that has surfaced afteer four decades of green revolution is not an easy one. There is no short cut. Several solutions have been suggested in the recent past. Most of these have one thing in common. Most of these, centered on organic path. Most agred that the new path should be a sustainable one. Sustainable solutions have been proposed in the name of:

- Natural farming.
- Organic farming.
- Regenerative agriculture.
- Low external input sustainable agriculture.
- Permaculture.
- Ecological agriculture.
- Integrated agriculture, etc.
  - Some said that the traditional agriculture is the answer.
  - Some said that replacement of inorganic inputs with organics will do.
  - Some said, let us reduce chemical inputs and also apply some organic inputs too because just the organic inputs alone cannot give enough productivity.
  - Others suggested structural and functional integration of farming components as road to sustainability.

The Concept

A farm is a system in that it has *inputs, processes andoutputs.* Depending on the type of farming (arable/pastoral, commercial/ subsistence), type and amount of inputs, processes and outputs will vary. Income through **arable fanning** alone is insufficient for bulk of the marginal farmers. Other activities such as dairying, poultry, sericulture, apiculture, fisheries, etc. assume critical importance in supplementing their farm income.

Utilisation of inputs without impairing the quality of environment is the main objective in the farming system approach. Therefore, it is clear *that farming system is a process in which sustainability of production is the objective.* The overall objective is to evolve technically feasible and economically viable farming system models by integrating cropping with allied enterprises for irrigated, rainfed, hilly and coastal areas with a view to generate employment and income from the farm.

Time and Space Concept

Time concept relates to increasing crop intensification in situation where there is no constraint for inputs.

In **rainfed areas** where there is no possibility of increasing the **intensity of cropping,** the other modem concept (space concept) can be applied. In **space concept,** crops are arranged in tier system combining two or more crops with varying field duration as **inter-crops** by suitably modifying the planting method.

Income through arable cropping alone is insufficient for bulk of the marginal farmers. Activities, such as dairy, poultry, fish culture, sericulture, bio-gas production, edible mushroom cultivation, agro-forestry and agri-horticulture, etc. assumes critical importance in supplementing farm income. It should fit well with farm level infrastructure and ensures full utilisation of by-products. **Integrated farming system** is only the answer to the problem of increasing food production for increasing income and for improving the nutrition of small-scale farmers with limited resources.

Farming system conceptually is a set of elements or components that are interrelated, which interact among themselves. At the centre of the interaction is the farmer exercising control and choice regarding the types of results of interaction. Income from cropping alone from small and marginal farm is insufficient now to sustain the farmers' family. A judicious mix of any one or more of these enterprises with agronomic crops should complement the farm income and help in **re-cycling** the farm residues/wastes. Selection of enterprises must be based on the cardinal principles of minimising the competition and maximising the **complementary** between the enterprises. Of late, the researchers on multi-disciplinary approach greatly realised and started developing the various farming system models in accordance with the agro-ecosystem zones.

Simmonds (1984) clarifies the farming system approach as follows: It is an academic activity comprising of theory, concepts, principles, approaches, etc. It creates an opportunity for developing diversified models for different type of farmers and different category of farmers. New fanning system approach models could be developed by means of on-farm research and extension. It causes consequential complex change which demands for Government interventions for farming system development.

Farming system is a problem solving approach for the farmer. It requires commonly homogeneous type of farmers. It is an inter-disciplinary approach. It is a participatory and bottom up planning. It requires on farm trials. It depends on the concept learning by doing and farming system approach needs socially desirable technologies.

Thus, the concept of farming system approach can be summarised as it is a holistic approach, complex in nature, inter-related of components, matrix of soils, plants, animals, power, implements, labour, capital and other inputs, influenced by political, economic, institutional and social forces.

1.5 Key Principles of Farming System

The three key principles of farming system are:

1. Cyclic.
2. Rational.
3. Ecologically sustainable.

Cyclic

The farming system is essentially cyclic (organic resources-livestock-land-crops). Therefore, management decisions related to one component may affect the others (Fig. 1.1).

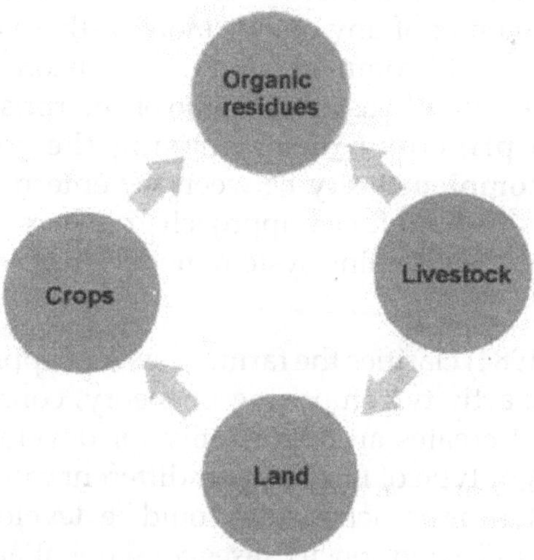

Fig 1.1 : Key principles of farming system

Rational

Using **crop residues** more rationally is an important route out of poverty. For resource- poor farmers, the correct management of crop residues, together with an optimal allocation of scarce resources, leads to sustainable production.

Ecologically Sustainable

Combining **ecological sustainability** and economic viability, the integrated livestock farming system maintains and improves agricultural productivity while also reducing negative environmental impacts.

1.6 Types of Farming System

Many farms have similarity in size, methods followed in cultural activities and outcome of products. When farms are quite similar in kind and production of the crops and livestock that are produced and methods and practices used in production, the group is called as **type of farming.** Farming system has been classified differently by different authors. One of the simple classifications is given in Table 1.1.

Table 1.1 : Different types of farming systems.

Arable, pastoral, mixed, organic	Subsistence or commercial	Intensive or extensive	Sedentary or nomadic
Amble is the growing of crops.	*Subsistence farmers* only produce enough food to survive. There is little, if any surplus left to sell.	*Intensive*-high inputs of money, labour or technology to achieve high outputs or yields per hectare. The farms are usually quite small.	*Sedentary* is when the settlement is permanent and the landscape farmed every year.
Pastoral is the keeping of animals. The only crops grown are fodder crops.	*Commercial farmers* sell their crops and animals in order to make a profit.	*Extensive-low* inputs, large areas of land, low outputs or yields per hectare.	*Nomadic* farmers move around looking for fresh pasture or new plots to cultivate.
Mixed is when farmers grow crops and rear animals.			
Organic farms are environmentally friendly. Chemicals and pesticides are not used. Animals are not reared intensively.			

On the basis of nature and purpose of farming, there are several types of farming systems in India such as subsistence farming, commercial farming, intensive farming, extensive farming, plantation farming, mixed farming and so on.

Subsistence farming: In subsistence farming, entire production is for family consumption. There is no surplus to sell in markets. Such type of farming is adopted by small and marginal farmers on fragmented land holdings. Cultivated crops, generally, are food crops and the methods of farming are generally archaic.

Commercial farming: It is opposite to subsistence farming. Its objective is to sell the produce in the market. Farming is, largely, practiced with modem tools and techniques, irrigation, chemical fertilisers, insecticides, pesticides and high yielding varieties of seeds. Cultivated crops include food crops commercial crops, fodder crops, etc.'

Extensive farming: Extensive farming is practiced on large patches of land. Output from such large swathes is huge but production per unit of land may be low. Such farming is common in developed countries (USA, Canada, etc.) with large farms. Extensive farming is almost absent in India except in some states, such as Punjab, Haryana and Uttar Pradesh.

Intensive farming: Intensive farming aims at maximum output per unit of land. It is common in India and south-east Asian countries such, as Thailand, Vietnam, Indonesia, etc.

Plantation farming: It is the farming in estates where single cash crop is grown for sale. Examples include crops, such as tea, coffee, rubber, banana, spices, etc.

Mixed farming: It refers to the livestock production along with usual cultivated crops. Farmers practicing mixed farming are economically better.

On the basis of the share of gross income received from different sources and comparative advantage, the farming systems of India can be classified as indicated below.

1.6.1 According to Size of Farm

According to size of the farm, farming systems are of two major types: collective farming and cultivation farming.

Collective Farming

It is direct collection of plant products from non-arable lands. It may include either regular or irregular harvesting of uncultivated plants. Honeying and fishing usually go hand in hand with collection. Actual cultivation is not needed. Natural products like honey, gum. flower, etc. are collected. Such plant products may be collected from forestry area.

Cultivation Farming

In this system, farming community cultivates the land for growing crops and rearing livestock.

Small Scale Farming

In this type, fanning is done on small size of holding and other factors of production are small in quantity and scale of production is also small.

Advantages

1. Intensive cultivation is possible.
2. Labour problem do not affect the production.
3. It is easy to manage the farm.
4. There is less loss due to natural calamities like frost, heavy rainfall and dis
5. Per unit output increases.

Disadvantages

1. Small scale farming cannot take advantages of various economic measures.
2. Cost of production per unit is more.
3. Mechanisation of agriculture is not possible.
4. Farmer does not get employment round the year.

Large Scale Farming

When farming is done on large size holding with large amount of capital, large labour force, large organisation and large risk is called large scale farming. In other words, when the factors of production are large in quantity, the kind of farming is said to be large. In India, 40 to 50 ha land holding may be said large scale farming but in countries like America, Canada and Australia even 100 ha farms are also called as small farms.

Advantages

1. Production is more economical as cost of production per unit is less.
2. Higher production per unit area.
3. Better marketing of agricultural products possible.
4. Processing, transportation, storage, packaging of produce is economical.
5. Costly machine like tractor, combined harvester can be maintained on the farm.
6. Subsidiary occupation, such as dairy, poultry, bee-keeping can be maintained on the farm.
7. Proper utilisation of factors of production is possible.
8. Research is possible.

Disadvantages

1. If demand for produce decreases and production exceeds the market demand, there will be more loss to the large farm.
2. Due to natural calamities like frost, drought, flood, insect pests and diseases or_ other influences the large farm will suffer a lot.
3. It will be difficult to manage large scale farm.

1.6.2 According to Land, Labour and Capital Investment

The farmer, on a given plot of land, obtains more or less definite quantity of production/ yield of commodity/commodities during any particular time. If he wants to increase his output, he can either: bring more land under cultivation or apply more labour and capital to the same piece of land.

Intensive Cultivation

In intensive cultivation, more labour and capital are used in the same piece of land. In other words, land remains fixed in quantity while other factors are increased. Intensive cultivation is preferred to extensive cultivation under the conditions of population pressure, low capital and cheap labour. In the earlier stages of development, population was small and technical knowledge of agriculture was also limited. Hence, extensive method was adopted. With increase in population, intensive cultivation become necessary and improvement in technology made its adoption possible.

Extensive Cultivation

When more area is brought under cultivation to increases the output, it is termed as extensive cultivation. In extensive cultivation, land is not a limiting factor of production but other factors of production (labour and capital) will be the limiting factors of production. In early times when land was plentiful, extensive cultivation was followed. Extensive and intensive cultivation go side by side in a country for a certain period of time and afterwards intensive cultivation may become more important. In most of the countries, extensive and intensive methods of cultivation generally go hand-in-hand.

1.6.3 According to Value of the Produce and Income

Based on the price offered to the produce or income from the produce, farming systems are of different types: specialised farming, diversified farming, mixed farming, ranching and dryland farming.

Specialised Farming

Farm in which 50 per cent or more income of total crop production is derived from a single crop is called specialised farming. The farm in which only single crop is cultivated for selling in the market and the income of the farm depends mainly on that crop is called specialised farming by Hopkins.

Advantages

1. *Better use of land:* More profitable to grow crops on land best suited to it (jute cultivation on swampy land in West Bengal).
2. *Better marketing:* It allows grading, processing, storing, transporting and financing the produce.
3. *Less equipment and labour needs:* Efficient use of equipment on specialised farm.
4. *Costly and efficient machinery can be used:* A wheat harvester/thresher can be maintained in a highly specialised wheat farm.
5. *Improvement in the efficiency and skill of the labour:* Specialisation allows a man to be more efficient and expert at doing a few things.

6. *Maintenance of farm records* will be easily.
7. *Better management:* Fewer enterprises on the farm are liable to be less neglect and sources of wastage can easily be detected.

Disadvantages

These disadvantages of specialisation are evident, when the farmer realises that "all eggs are in one basket".

1. *Greater risk:* Crop failure or decreasing market price of the produce is the maj risks with specialised farming.
2. *Loss of soil fertility and productivity* will be at risk due to specialised farming.
3. *In efficient use of land, labour and capital.*
4. *Unevenly distributed farm income per unit time* (season, year).
5. *Under-utilisation of crop growth resources.*
6. Specialised farming does not meet all the food needs of the family members.

Mixed Farming

In mixed farming crop production is combined with the rearing of livestock. Livestock enterprises (cows, buffaloes, sheep, goat and fisheries) are complementary to crop production. In mixed farming, at least 10 per cent of its gross income must be from livestock activity. The upper limit is 49 per cent under Indian conditions. In mixed farming, cow and buffaloes are included with crop production.

Advantages

1. It offers highest return on farm business.
2. It provides work throughout year.
3. Efficient utilisation of land, labour, equipment and other resources.
4. Crop by-products, such as straw, fodder, etc. are used for feeding livestock.
5. Farm yard manure from livestock maintains soil fertility.
6. It meets food needs of the family members.
7. Intensive cultivation is possible.

8. If one source of income is lost, he can maintain his family from other source of income.
9. Milk cattle provide draft animals for crop production and rural transport.
10. Mixed farming increases social status of the farmer.

In India, livestock is much closely connected with agriculture because animal power is the main source of power in agriculture. FYM is the main source for maintaining soil fertility and animals make good use of subsidiary and by-products on farms and in turn they provide milk. Mixed farming appears ideal under Indian conditions.

Disadvantages

1. It is still a primitive method of crop production in India.
2. Maintenance of different components in the system is a problem.
3. Labour requirement is more in this farming.

Ranching

A ranch differs from crop and livestock farming in that the livestock graze the natural vegetation. Ranch land is not utilised for tilling or raising crops. Ranchers have no land of their own and make use of the public grazing land. A rancher occupies most of the time of one or more operators. Ranching is followed in Australia, America, Tibet and certain parts of India.

Dryland Farming

Farmers of dryland (around 750 mm rainfall) are subjected to loss due to vagaries of monsoon. Dryland farming needs attention on the following aspects of crop production:

1. Timely ideal seedbed preparation for soil and water conservation and optimum crop stand establishment.
2. Timely inter cultivation to minimise adverse effect of weeds.
3. Improving water retentive capacity of soil through application of organic manures.
4. Adoption of soil and water conservation practices.
5. Use of improved cultivars.
6. Maintenance of soil health.

Environmentally sustainable dryland farming system emphasises conservation and utilisation of natural resources. Agronomic practices of conservation, tillage and **mulch farming, rotational cropping,** use of legumes and cover crops for improving **soil fertility** and suppressing weeds and efficient uses of cattle manure are some of the components of sustainable farming system.

1.6.4 According to Type of Rotation

The word rotation has two meanings according to the time period involved. There is long term alteration between various types of land use, such as **arable farming, tree farming,** grassland use, etc. In this, rotation means the sequence of this basic type of land use on a given field. Within arable farming, there is also the term **crop rotation** which means the short-term sequences of different arable crops on one field.

Lay System

In this system, several years of arable farming are followed by several years of grassed and legumes utilised for livestock production.

Unregulated lay system: In this system, natural vegetation grasses, bushy growth on pasture are allowed to grow during the period of fallow. This is an improved managed pasture.

Regulated lay system: During the period of **fallow,** certain types of grasses are grown or planted. These are the well-managed pasture with fencing and adopting rotational grazing system.

Perennial Crop System

The crops which cover the land for many years (tea, coffee, sugarcane, etc.) are called as perennial crops. In some cases, tree crops (oil palm, rubber) are alternated with **fallow** or with other **arable farming,** grazing, etc.

Farming System According to Intensity of Rotation

Intensity of the rotation denoted by *"R"* is given by:

$$R = \frac{\text{Number of years of cultivation}}{\text{Length of the cycle of land utilisation}} \times 100$$

Length of the cycle is the sum of the number of years of arable farming. Number of fallow years "/?" indicates the production of the area under cultivation in relation to a total area available for arable farming.

- *Shifting cultivation* **(R < 33%)**: There is more number of years of fallow than actual cultivation.
- ***Lay or fallow farming:*** In this case, *R* < 66% and 33%.
- ***Permanent cultivation (R> 66%):*** In this case, large area is cultivated and small area is left fallow.
- ***Multiple cropping (R = > 100%):*** *R* is 150 per cent means, 50 per cent area is under two crops in a year. *R* is 300 per cent means; three crops are grown in an year.

1.6.5 According to Degree of Commercialisation

Depending upon the produce sold in the market for earning money, farming systems are grouped as indicated below:

Commercial Farming

More than 50 per cent of the produce is for sale.

Partly Commercial Farming

More than 50 per cent of the value of produce is for home consumption.

Subsistence Farming

Virtually there is no sale of crop and animal products. Entire produce is used for home consumption. Subsistence farming is a type of farming where the farmers of our country cultivate the crop in their land for the livings. Hence, the holdings are small in size. No scope for improved cultivation practices. Such farmers usually resort to raring cattle, poultry, along with crop cultivation in limited land to meet their domestic requirement.

Advantages

1. Efficient use of productive resources.
2. All the family members engaged in farm activities, such as cop cultivation, cattle rearing, poultry, etc. through the year.

3. Farmer meets their demand from the income due to cattle raring, poultry, etc.
4. Efficient use of farm by-products.

Disadvantages

1. Small holding may not permit adaption of improved farming practices.
2. Crop production is limited to monsoon rains.
3. Income from subsistence farming is very low.

1.6.6 According To Degree Of Nomadic, Cropping Pattern And Implements

Grassland Farming

This type of farming is associated with rearing of animal for economic production. It is classified on the basis of nomadic:

Total nomadic: In this system, the animal owners do not have permanent place of residence. They do not practice regular cultivation. Their families move with the herds.

Semi-nomadic: Animal owners have permanent place of residence, while supplementary cultivation is practiced. However, for long periods of time, they travel their herd to distant grazing areas.

Transhumant: Seasonal migration of livestock to suitable grazing ground or it is the situation in which farmer with a permanent residence sends their herd with herdsman for long period of time to distant grazing areas.

Partial nomadic: Farmer has permanent residence and who have herds at their disposal, which remains in the vicinity.

Stationary animal husbandry: The animals remain on the holding or in the village throughout the entire year.

Major farm systems of India are dryland and irrigated system. Discussion on these farming systems is beyond the scope of this publication.

There are six basic farm types in India and surrounding countries:

- **Type 1.** Small subsistence-oriented family farms.
- **Type 2.** Small semi-subsistence or part-commercial family farms with 1 to 2 ha. The area is not a good criterion. The same basic

structure can be found on much larger 20 to 30 ha farms as in the Punjab, Sindh and north-west frontier provinces of Pakistan.

- **Type 3.** Small independent specialised family farms.
- **Type 4.** Small dependent specialised family farms, often with the family as tenants.
- **Type 5.** Large commercial family farms, usually specialised and operated along modified estate lines.
- **Type 6.** Commercial estates, usually monocrop and with hired management and absentee ownership.

1.6.7 Indigenous Farming System

Shifting cultivation, *taungya* cultivation and *zabo* cultivation are the three major indigenous farm systems in India.

Shifting Cultivation

It refers to farming system in north-eastern areas in which land under natural vegetation (usually forests) is cleared by slash and bum method, cropped with common **arable crops** for a few years and then left unattended when natural vegetation regenerates. Traditionally, the fallow period is 10-20 years but in recent times, it is reduced to 2-5 years in many areas. Due to increasing population pressure, the fallow period is drastically reduced and system has degenerated causing serious soil erosion, depleting soil fertility resulting to low productivity. In north-eastern India, many annual and perennial crops with diverse growth habits are being grown.

Taungya Cultivation

The *Taungya* system is like an organised and scientifically managed **shifting cultivation.** The word is reported to have originated in Myanmar (Burma) and *tauang* means hill, *ya* means cultivation *i.e.,* **hill cultivation.** It involves cultivation of crops in forests or forest trees in crop-fields and was introduced to Chittagong and Bengal areas in colonial India in 1890. Later, it had spread throughout Asia, Africa and Latin America. Essentially, the system consists of growing annual arable crops along with the forestry species during early years of establishment of the forest plantation. The land belongs to forest department or their large scale leases, who allow the subsistence farmers to raise their crops and in turn protect tree saplings. It is not merely temporary use of a piece of land and a poverty level wage, but is a chance to participate equitably in a diversified and sustainable agro-forestry economy.

Zabo Cultivation

The *Zabo* is an indigenous farming system practiced in north-eastern hill regions, particularly in Nagaland. This system refers to combination of forest, agriculture, livestock and fisheries with well-founded soil and water conservation base. Rainwater is collected from the catchment of protected hill tops of above 100 per cent slopes in a pond with seepage control. Silt retention tanks are constructed at several points before the runoff water enters in the pond. Cultivation fully depends on the amount of water stored in the pond. Land is primarily utilised for rice. This system is, generally, practiced in high altitude hill areas, where it is not possible to construct terraces and or irrigation channels across the slope. This is a unique farming system for food production to make livelihood. *Zabo* means impounding of water. The place of origin of *zabo* farming system is thought to be the Kikruma village in Phek district of Nagaland.

1.7 Factors Affecting Types of Farming

Functioning of any individual farming system is strongly influenced (determined) by the external rural environment, including policies and institutions, markets and information linkages. Regardless of their size, individual farming systems are organised to produce food and to meet other household goals through the management of available resources—whether owned, rented or jointly managed—within the existing social, economic and institutional environment (FAO 2001). They often consists of a range of interdependent activities *viz.*, crop production, post-harvest processes, livestock keeping, fishery, agro-forestry, etc. Off- farm incomes, which make a significant contribution to the livelihood of many rural poor, are also included. Systematic representation of the different exogenous and endogenous factors influencing the development of farming system is given in Fig 1.2.

This figure represents the inter-relationship between the key determinants of farming systems. Some of these factors are internal to or part of the farming system, whereas others are external. Principal exogenous (external) factors influencing the development of farming system include: policies, institutions, public goods, markets and information are indicated on the left side of the figure, lying outside the dotted line that marks the system boundary. Availability of markets and the prices influence a farmer's decision on enterprise pattern, on purchases of inputs and on the timing of the produce sales. Availability of economic social infrastructure in rural areas determines the transport costs and the availability of services to the household, notably human and animal health. Similarly, information and educational services affect household strategy and decisions. Technologies which determine the nature of production and processing and natural resources are largely endogenous (internal) factors and are, therefore, depicted as lying mainly within the boundary of the farming system. In general terms, the biophysical factors tend to define the set of possible farming systems, whilst the socio-economic factors determine the actual farming system which can be observed at a given time.

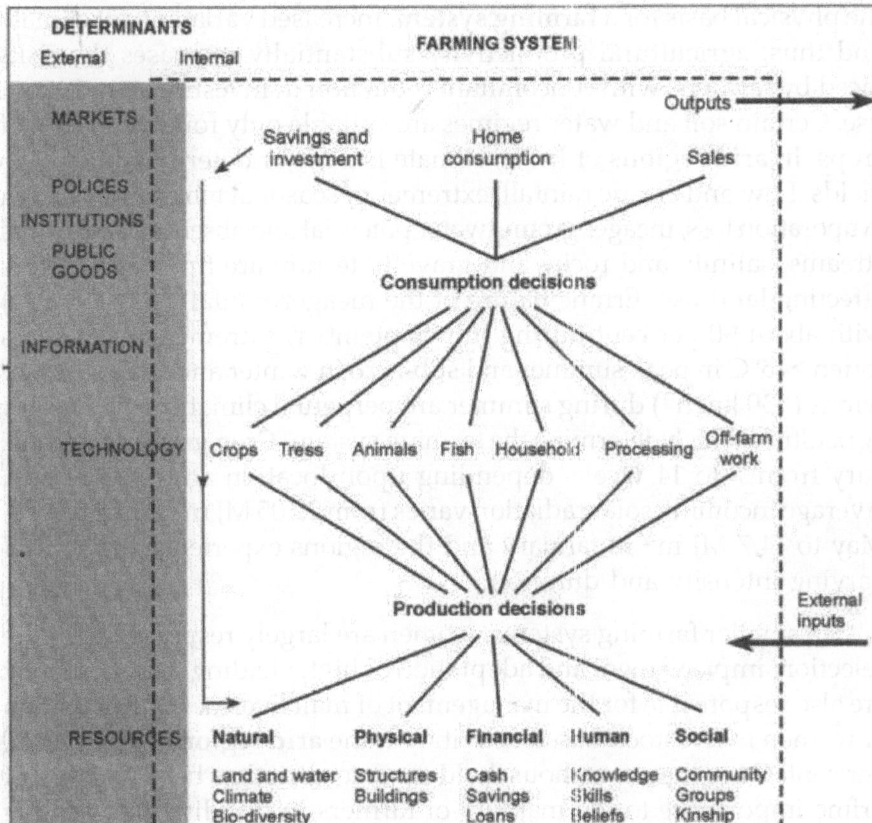

Fig. 1.2 : Determinants of farming system. Dotted lines show the boundary between internal and external factors (FAO 2001).

Functioning of any individual **farming system** is strongly influenced by the external rural environment. Not only farms are closely linked to the off-farm economy through commodity and labour markets, but the rural and urban economics are also strongly interdependent. For example, as noted above, it is quite common for small farm households to derive a significant part of their income, often 40 per cent or more, from off-farm activities. Farm men and women are also linked to rural communities and social networks and this social capital influences the management of farms. The key categories of determinants influencing a farming system are as follows:

1.7.1. Natural Resources and Climate

Interaction of natural resources, climate and population determines the physical basis for a **farming system.** Increased variability of climate and thus, agricultural productivity substantially increases the risks faced by farmers, with concomitant reduction in investment and input use. Certain soil and water regimes are suitable only for given types of crops. In **arid regions** of India, climate is a major determinant of crop yields. Low and erratic rainfall, extremes of seasonal temperature, high evaporation loss, meager groundwater potential and absence of perennial streams, salinity and rocky and gravelly terrain are the major factors affecting land use. Erratic nature of the meager rainfall (100-450 mm) with about 90 per cent during July-September, extreme temperatures (often >45°C in peak summer and sub-zero in winter and high summer winds (>30 km h^{-1}) during summer are perpetual climatic problems for agriculture which determine the farming system. **Crop growing periods** vary from 7 to 14 weeks depending upon location and type of soil. Average incoming solar radiation varies from 22.05 MJ m^{-2} during April-May to 14.7 MJ m^{-2} in January and the regions experience drought of varying intensity and duration.

In smaller farming systems, women are largely responsible for the selection, improvement and adaptation of high yielding varieties. They are also responsible for the management of mall livestock. Participation of women in livestock-based activities in the **arid regions** was up to 90 per cent. Over the years, household food security has been an issue of prime importance to the majority of farmers in deciding the farming system. Similarly, physical and geographical features (drainage characteristics, elevations and slopes), climatic factors (total rainfall and its distribution, minimum and maximum temperature, humidity and intensity of sunfight) are other factors which have to be taken into consideration when making decisions with respect to selection of enterprise for a farming system.

1.7.2 Science and Technology

Appropriate technology for different agroclimatic conditions and local needs of the people helps determine the farming system of a region. Investment in agricultural science and technology has expanded rapidly during the last 4-5 decades. During this period, major technical and institutional reforms have occurred, which have shaped the pattern of technological development and dissemination. Research-driven growth in developing countries has largely been a result of the **green revolution,** which has resulted in considerable advance a die field of food grain production, with the support of policy and other aspects. Research has been focussed, principally, upon intensifying crop and livestock production. There has been much less research on integrated technologies for diversifying the livelihoods of small farmers in developing countries and increasing the sustainability of land use.

1.7.2. Trade Liberlisation and Market Development

Marketing and processing are the two basic pre-requisites for promoting IFS under given agro-ecological conditions. Once the market for alternative produce from IFS is assured, farmers get motivated to diversify their farming system. Similarly, the rural marketing system plays a key role in agriculture diversification. As a result of reduction of impediments to international trade and investment, the process of trade liberalisation is already generating changes in the structure of production at all levels including smallholder-farming systems in many developing countries. Not only the market development is accelerating, but patterns of production and natural resources usage are also changing profoundly in response to market forces. Availability of new production, post-harvest and transport technologies will also change demand patterns by making possible the delivery of new products or established products in new forms to markets, where they have been previously unattainable. With the World Trade Act coming in to force, Indian agriculture is more exposed to the global market than ever before.

1.7.4. Policies, Institutions and Public Good

Policy frameworks of government for rural infrastructure transport, irrigation, storage facilities, processing and providing incentives to the farmers encourages agricultural diversification.

Development of a dynamic farming system requires a policy environment which is conducive. Moreover, the establishment of farm-rural-urban linkages requires effective demand. Policy-makers have increasingly shifted their attention to the potential to increase the efficiency of service delivery through restructuring of institutions. Production incentives have a dramatic effect on farming system. Policies on land ownership, water management and taxation reform, *inter alia*, have a great bearing on the types of farming system in a region or area. Efficient farming system needs efficient energy management.

1.7.5 Information and Human Capital

Evolution of a farming system based on increasing specialisation (large scale broiler units) or integrated intensification (rice-fishery-duckery) requires extra knowledge on the part of farm operators. Need for better information and enhanced human capital has also increased, as production systems have become more integrated with regional, national and international market systems. Lack of education, information and training is frequently a key limiting factor to smallholder development. Many observers anticipate that an information revolution, that is the bridging of the knowledge gap between scientists and farmers, will be a key to agricultural growth of these small farms. It is increasingly recognised and acknowledged by development workers that the empowerment of women is crucial to raising levels of child and family nutrition, improving the production and distribution of food and agricultural products and enhancing the living conditions of rural populations. Similarly, better access to credit, land and extension services would enable women to make an even greater contribution to eliminating rural hunger and poverty.

1.7.6 Indigenous Technological Knowledge

Indigenous technical knowledge (ITK) is that which people in a given community have developed over time and continue to develop. It is based on experience, often tested over long periods of use, adapted to local culture and environment, dynamic and changing and lays emphasis on minimising risks rather than maximising profits. It covers a wide spectrum: soil, water and nutrient management, pasture and fodder management, crop cultivation, plant protection, farm equipment, farm power, post-harvest preservation and management, agro- forestry, bio-diversity conservation and also exploitation, animal rearing and health care, animal products preservation and management, fisheries and fish preservation and ethnic foods and homestead management. Thus, the indigenous technical knowledge of a farmer has a great influence in managing the farm and farming system. During the first half of the 20th century, when there was practically no external source of technological change, indigenous knowledge dominated the farming system and agricultural output grew less than 0.1 per cent per annum and that too was primarily on account of systematic expansion of labour and land.

1.8 Farming System Components

Under the All India Coordinated Project on IFS, survey of existing fanning systems in 732 marginal households in the country spread over 30 districts in 20 states reveled the existence of 35 farming systems with as many as components. Out of the sample surveyed, crop-livestock and crop-dairy systems were practiced by around 50 per cent of marginal farmers followed by crop-dairy-goat (10%). Among the livestock category, dairy accounts for 86 per cent of the farmers followed by goat (24%) and poultry (21%). On any given farm, farming subsystems or farming activities may comprise any one of the combination of the following components (crop production, livestock, poultry, apiculture, sericulture, mushroom cultivation, aquaculture, agro-forestry).

1.8.1 Crop Production

Every farmer concentrates on crop production to feed the family. In India, food crop is predominantly grown in most ideal season and thus a particular food crop is basic to the cropping system. Accordingly, cropping systems are referred to as: rice-based, sorghum- based, pearlmillet-based, wheat-based, groundnut-based, sugarcane-based, vegetable-based, etc. The practice of growing different crops of varying heights and duration is called multi- storeyd/multi-tire cropping.

1.8.2 Livestock (Milk and Draught Animals)

For centuries farmers of India are growing crops and livestock together and livestock has been the integral part of the system. Apart from contributing to food basket, nutritional security and household income of the farmers, this system plays a significant role in generating employment to landless, small and marginal farmers and farm women.

1.8.3 Poultry

Rearing poultry in a backyard system is an age old practice. Backyard poultry is a means of livelihood for a fairly large number of small and marginal farmers in India. Poultry in IFS plays a pivotal role in fulfilling the ever increasing demand for meat and eggs in the country.

1.8.4 Apiculture

This is the science and culture of honeybees and their management. It requires low investment and so, it can be taken up by small, marginal and landless farmers and the educated unemployed youth. The honey bee species Apis cerana indica and A mellifera are very commonly used in India to the economic benefit of small and marginal farmers.

1.8.5 Sericulture

This is defined as the practice of combining mulberry cultivation, silkworm rearing and silk reeling. India occupies second position among silk producing countries in the world, next to China. Sericulture is labour intensive in all its phases, including off-farm activities, such as twisting, dyeing, weaving and printing. It has considerable socio-economic importance in India, largely due to its suitability for small and marginal farm-holdings by generating, employment and requiring low investment.

1.8.6 Mushroom Cultivation

One of the important areas for using agro-wastes to enhance income and impart a higher level of sustainability to the system is mushroom cultivation. In Indian situations, mushroom is considered as an important component of farming systems owing to its need for less space, better year-round employment and income generation potential and suitability for small and marginal holdings.

1.8.7 Aquaculture

This is considered as very promising and stable component of farming system in the coastal belt of India and in other Asian countries, such as Bangladesh, Vietnam, China, Indonesia and Malaysia. Aquaculture has the potential to improve sustainability and the income generation of small farmers, when fully integrated with other enterprises and household activities so as to allow farm families and communities to manage their natural resources effectively.

1.8.8 Agro-forestry

This provides a land use option that increases livelihood security and reduces vulnerability to climate and environmental change. According to the planning commission report Greening India, 33 per cent forest cover can only be achieved through agro-forestry. Agro-forestry is a valuable component of farming system that has many potential advantages (biomass productivity enhancement, soil fertility improvement, soil conservation, nutrient cycling, microclimate improvement, carbon sequestration, bio-drainage, bio-energy, and bio-fuel production).

Other location specific farming system component include: goat and sheep rearing, piggery, duckery, fishery, biogas, etc.

1.9 Terms Used in Cropping Systems

Crop intensification techniques include inter-cropping, sequence cropping, relay cropping, etc. All such terms come under the general term multiple cropping. To date, there is no complete agreement on the terms used in cropping systems. The first organised attempt to provide guidelines based on the options of a wide range of specialists occurred in the symposium on multiple cropping in 1975 at the annual meeting of the American Society of Agronomy in Knoxville, Tennessee. Results of consensus at that meeting were summarised by Andrews and Kassam (1976).

Cropping System and Pattern

Cropping pattern and cropping system are two terms used interchangeably; however these are two different concepts. While cropping pattern refers to the yearly sequence and spatial arrangement of crops or of crops and fallow in a particular land area, cropping system refers to cropping pattern as well as its interaction with resources, technology, environment, etc.

Thus, a cropping system comprises cropping pattern plus all components required for the production of a particular crop and the inter-relationships between them and environment. Cropping system is a critical aspect in developing an effective ecological farming system to manage and organise crops so that they best utilise the available resources (soil, air, sunlight, water, labour, equipments). It represents cropping patterns used on a farm and their interaction with farm resources and farm enterprises and available technology which determine their makeup. It is executed in the field level. Some other terms, frequently, used in crop production systems are given below:

Agro-ecology: The study of the relation of agricultural crops and environment.

Agro-forestry: It is a self-sustaining land management system which combines production of agricultural crops with tree crops and also with livestock simultaneously or sequentially on the same unit of land.

Alley cropping: A farming system in which arable crops are grown in alleys formed by trees or shrubs, established mainly to hasten soil fertility re-storation and enhance soil productivity and for shelter.

Calorie index: A single productivity index which incorporates caloric equivalents produced by all the crops.

Competition index: Competition of intercropped species for light, nutrients, water, carbon dioxide and other growth factors.

Complementary effect: Effect of one component on another which enhance growth and productivity, as compared to competition.

Component crops: Individual crop species which are a part of the multiple cropping systems. Crop rotation: Repetitive cultivation of an ordered succession of crops (or crops and fallow) on the same land. One cycle often takes several years to complete.

Cropping index: The number of crops grown per annum on a given area of land multiplied by hundred.

Double cropping: Also known as sequential cropping is the practice of planting a second crop immediately following the harvest of a first crop, thus harvesting two crops from the same field in one year. This is a case of multiple cropping, which requires crops that mature quickly enough to allow two harvests in one year.

Ecofarming: It is the potential for introducing mutually reinforcing ecological approaches to food production. It aims at the maintenance of soil chemically, biologically and physically the way nature would do, if left alone. "Feed the soil, not the plant" is the watchword and slogan of ecological farming.

Intercropping: It refers to growing of two or more, generally, dissimilar crops simultaneously on the same piece of land. Usually, the base crop is grown in a distinct row arrangement. The recommended optimum plant population of the base crop is suitably combined with appropriate additional plant density of the associated crop and there is crop intensification in both time and space dimensions.

Mixed cropping: Growing of two or more crops simultaneously on the same piece of land, without any definite row arrangement.

Mixed intercropping: Growing two or more crops simultaneously with no distinct row arrangement.

Monetary index: A single productivity index which incorporates monetary value of all the crops in a system.

Monocropping or monoculture: It refers to the presence of a single crop in a field. This term is often used to refer to growing the same crop year after year in the same field. This practice is better described as continuous cropping or continuous monocropping.

Multiple cropping: Growing two or more crops consecutively or at the same time on the same field in the same year.

Multi-story/multi-tier cropping: It is a system of growing together, crops of different heights at the same time on the same piece of land and thus, using land, water and space most efficiently and economically (coconut + pepper + pineapple + grass).

Relay inter-cropping: It is a technique in which different crops are planted at different times in the same field and both (or all) crops spend at least part of their season growing together in the field. An example would be dropping cover-crop seed into a soybean crop before it is mature.

Row intercropping: Growing two or more crops simultaneously where one or more crops are planted in rows.

Sole cropping: One crop variety grown in pure stand at normal density.

Spatial arrangement: The physical or spatial organisation of component crops in a multiple cropping system.

Stripcropping: Growing soil conserving and soil depleting crops in alternate strips running perpendicular to the slope of the land or to the direction of prevailing winds for the purpose of reducing erosion.

1.10 Management of Crop Production Components

The potential enterprises which are important in farming system in the way of making a significant impact of farm by generating adequate income and employment and providing livelihood security are briefly presented.

Crop production is an integral part of farm activities in the country. Cropping system based on climate, soil and water availability have to be evolved for realising the potential production levels through efficient use of available resources. The cropping system should provide enough food for the family, fodder to the cattle and generate sufficient cash income for domestic and cultivation expenses. These objectives could be achieved by adopting intensive cropping. Methods of intensive cropping include multiple cropping and inter-cropping. Intensive cropping may pose some practical difficulties, such as shorter tum-around time lapse for land preparation before the succeeding crop and labour shortage at peak periods of agricultural activities. These practical handicaps can easily be overcome by making modifications in the cropping techniques. Alteration of crop geometry may help to accommodate inter-crops without losing the base crop population. Management of crop production components has been discussed in terms of sequential cropping systems and multi-tier cropping systems.

1.10.1 Management of Sequential Cropping System

In sequential cropping, the preceding crop has considerable influence on the succeeding crop. This includes the complementary effects such as release of N from the residues of the previous crop, particularly legume, to the following crops and carries over effects of fertiliser applied to preceding crops. The adverse effects include allelopathy, temporary immobilisation of N due to wide C : N ratio of the residues and carry over effect of pest and diseases.

In India, food crop is, predominantly, grown in most suitable seasons and thus particular food crop is basic to the cropping system followed by the farmers. Accordingly, the cropping systems are, usually, referred to as:

- Rice-based cropping system.
- Sorghum-based cropping system.
- Pearl-millet-based cropping system.
- Wheat and gram-based cropping system.
- Maize-based cropping system, etc.

Some of the cropping systems based on commercial crops are:

- Cotton-based.
- Groundnut-based.
- Sugarcane-based.
- Plantation crop-based.
- Vegetable-based cropping systems.

Seedbed preparation: Two problems are encountered in seed-bed preparation in sequential cropping systems:

1. The time available for seed-bed preparation is less in **high intensity cropping** system. Rains very frequently interfere with land preparation.
2. Due to the effect of the previous crop, the field may not be in proper condition to carry out field operations.

Water management: Water management for sequential cropping is same as for individual crops. There are no carry over effects of irrigation as in the case of fertiliser.

Weed management: Weed management practices are followed for the entire system as a unit instead of considering weed problem of individual crops. Herbicides applied to the previous crop may be toxic to the succeeding crop. Higher dose of atrazine applied to sorghum crop affects germination of succeeding pulse crop. Recommendation of **herbicides** and its dose should, therefore, take into consideration the succeeding crop.

Pests and diseases: Infestation of pests and diseases are more in **sequence cropping** system due to continuous cropping. Carry over effects of insecticides are not observed so far.

Harvesting: In sequence cropping, the crop can be harvested at **physiological maturity** instead of at **harvest maturity.** Field can then be vacated one week earlier for planting another crop. Because of continuous cropping, the harvesting time may coincide with heavy rains and special post-harvest operations like artificial drying, treating the crop with common salt, etc. are practiced to save the produce.

Grain production potential in different regions of the country under **intensive cropping** ranges from 11-18 t ha^{-1} in maize-potato or toria-wheat-moong system followed at IARI, New Delhi. It was possible to produce 14-15 tonnes of food per ha per annum without impairing the soil health. Results of **multiple cropping** demonstrations under irrigated conditions showed that production potential can be as high as 19.8 t ha^{-1} in cereal-based cropping system of rice-rice-rice. Yield potential of multiple cropping varies from region to region depending upon the physical and socio-economic conditions.

1.10.2 Management of Intercropping System

Crops are grown simultaneously. Management practices should, therefore, aim to provide favourable environment to all the components, exploit favourable interactions among the component crops and minimise competition among the components .Prominent interactions in **intercropping** include:

Light: Intercropping can increase light inception by as much as 30-40 per cent. Proper choice of crops and varieties, adjustment of planting density and pattern are the techniques to reduce competition and increase the **light use efficiency.**

Moisture and nutrients: Competition for water and nutrients results in two main types of effects on the less successful or suppressed component. First, the roots of dominated crop may grow less on the sides of aggressive component. Secondly, plants affected by competition for soil factors are likely to have increased root/shoot ratio.

Allelopathy: Some crops may be unsuitable to be grown as intercrops because they may produce and excrete toxins into the soil which are harmful to other components.

Annidation: Refers to complementary interaction which occurs both in space and time.

Crops are grown simultaneously. Management practices should, therefore, aim to provide favourable environment to all the components, exploit favourable interactions among the component crops and minimise competition among the components.

Seed-bed preparation: Seed-bed preparation is, generally, done as per the needs of base crop.

- Deep rooted crops respond to deep ploughing while for most of the cereals shallow tillage is sufficient.
- Crops with small seed require fine seed-bed.
- Certain crops like cotton and maize are planted on ridges, while most of the other crops are grown on flat seed-bed.

- Seed-bed for sugarcane, as usual, is made into ridges and furrows. Sugarcane is planted in furrows and inter-crops are sown on ridges.
- In groundnut + pigeonpea, flat seedbed is preferred for sowing the crops.
- **ICRISAT** is recommending **broad bed and furrows** for black soils of semi-arid regions for pure crops as well as inter-crops grown under rainfed conditions.
- Where the crop requirements are quite different as in rice + maize under rainfed conditions and also in agro-forestry, seedbed preparation forestry, is done separately for component crops. In rice + maize **inter-cropping** system, ridges and trenches are formed. Maize is planted on ridges and rice in trenches.
- In agro-forestry, pits are dug for tree species and a rough seed-bed is prepared in interspaces for the introduction of forage crops.

Varieties: Recommended principals should be followed for realising the benefits of inter-cropping:

- Varieties of component crops should be less competing with the base crop and the peak nutrient demand period should be different from the base crop.
- Difference in duration between the components in inter-cropping should be a minimum period of 30 days (maize + soybean, sorghum + red gram, *toria* + *gobhi sarson*).
- Selection of compatible genotypes of component crops increases the complementarity of inter-cropping system.
- Varieties selected for inter-crop should have thin leaves, tolerant to shading and less branching since these crops are, generally, shaded by the base crop.
- If the base crop is shorter than inter-crop, the inter-crop should be compact with erect branching and its early growth should be slow.
- Characteristics of base crop should be as in sole crop.

1.10.3 Nutrient Management

When legumes are associated with cereal crop in inter-cropping system, a portion of nitrogen requirement of cereal is supplemented by the legume. The amount may be as small as a few kg to 20 kg ha^{-1}. Cereal+legume inter-cropping, is therefore; mainly advantageous under low fertiliser application. Considering all the factors, it is suggested that the nitrogen dose recommended for **base crop** as pure crop is sufficient for inter-cropping system with cereals+legume or legume+ legume. With regard to phosphorus and potassium, one-eighth to one-fourth of the recommended dose of intercrop is also added in addition to recommended dose of base crop to meet the extra demand. Basal dose of nitrogen is applied to rows of both components in cereals+ legume inter-crop system. Top dressing of nitrogen is done only to cereal rows. Phosphorus and potassium are applied as basal dose to both crops.

1.10.4 Water Mangement

Inter-cropping systems are, generally, recommended for **rainfed crops** to get stable yields. Total water used in inter-cropping system is almost the same as for sole crops, but yields are increased. Thus, **water-use efficiency** of inter-cropping is higher than sole crops. Component crops differ on their capacity to withstand excess or difficult moisture conditions. However, the irrigation schedule followed for sole crops is suitable even for inter-cropping system. Scheduling irrigation at **IW/ CPE ratio** of 0.6 to 0.8 or irrigation at one bar soil moisture tension is suitable for most of the system. However, information on this aspect is meager.

1.10.4 Weed Management

Weed problem is less in inter-cropping system compared to their sole crops. In certain situations, inter-crops are used as biological agents to control weeds. Blackgram, greengram, cowpea in sorghum and cowpea in banana reduce weed population and one hand weeding can be avoided by this method. In some inter-cropping system like maize + groundnut, rice + tapioca, maize + tapioca, weed problem is similar to their sole crops. Growth habit of genotype used in inter-cropping has a great influence on weed growth. Weeds present in sole crops are different than those present in inter-cropping system. Though weed problem is less, weed control measures are necessary in inter-cropping system. But the labour required for weeding is less. Second weeding is not necessary because of crop coverage. Chemical weed control is difficult in intercropping system because the herbicide may be selective to one crop, but non-selective to another.

1.10.6 Pests and Diseases Management

Pests and diseases are believed to be less in inter-cropping system due to crop diversity than in sole crops. Some plant combinations may enhance soil **fungistasis** and **antibiosis** through indirect effects on soil organic matter content. Spread of the disease is altered by the presence of different crops. Little leaf of brinjal is less when brinjal is sheltered by maize or sorghum. As the insect carrying virus first attacks maize or sorghum, virus infection is less on brinjal. Non-host plants in mixtures may emit chemicals or odours that affect the pests, thereby protecting host plants.

1.10.7 Multi-Tier Cropping Management

The practice of growing different crops of varying height, rooting pattern and duration is called 'multi-tier cropping' or **multi-storeyed cropping.** Multi-storeyed cropping is mostly prevalent in plantation crops like coconut and areca nut. There is scope for **inter-cropping** in coconut garden up to the age of 8 years and after 25 years. During this period there is adequate light transmission to the ground, which permits cultivation of intercrops. The objective of this system of cropping is to utilise the vertical space more effectively. In this system, the leaf canopies of intercrop components occupy different vertical layers. The tallest components have foliage tolerant of strong light and high evaporative demand and the shorter component(s) with foliage requiring shade and on relatively high humidity (coconut + black pepper + cocoa + pineapple).

In this system, coconut is planted with a spacing of 7.5 m. Rooted cuttings of black pepper are planted on either side of coconut about 75 cm away from the base. On the coconut trunk at a height of about one metre from the ground level, the vines of pepper are trailed. A single row of cacao is planted at the centre of space between coconut rows. Pineapple is planted in the interspaces.

Coconut growing to a height of more than 10 m occupies the top floor. Black pepper growing to about 6-8 m height forms the second floor. Cacao with its pruned canopy of about 2.5 m height and pine apple growing to about 1 m height form the first and ground floors, respectively.

In another multi-tier system in coconut, ginger or turmeric and partial shade loving vegetables form the first tier, banana in second tier, pepper in third tier and coconut or areca nut in the final tier.

In the areca nut plantation, tuber crops are pre-dominantly inter-cropped. Elephant yam, tapioca, greater yam and sweet potato are common intercrops in humid tropics. Banana and pine apple are also cultivated as intercrops in areca nut gardens.

In coffee based multi-tier cropping system, first tier is with pine apple, second tier with coffee, third tier with cacao/mandarin and final tier with fast growing shade trees necessary for coffee plantation (dadaps and silver oak).

1.11 Efficient Cropping Systems and thier Evaluation

Appropriate decisions on ideal cropping systems under diversified **agro-ecological** situations depend on the productivity of the system under different situations. Productivity of the tested systems has to be evaluated to take final decision in this regard.

1.11.1 Efficient Cropping Systems

Efficient cropping system for a particular farm depends on farm resources, farm enterprises and farm technology because farm is an organised economical unit. Farm resources include land, labour, water, capital and infrastructure. When land is limited, **intensive cropping** is adapted to fully utilise available water and labour. When sufficient and cheap labour is available, vegetable crops are also included in the **cropping system** as they require more labour. Capital intensive crop like sugarcane, banana, turmeric, etc. find a space in the cropping system when capital is not a constraint. In low rainfall regions (750 mm per annum) **monocropping** is followed and when rainfall is more than 750 mm, **inter cropping** is practiced. With sufficient irrigation water, triple and **quadruple cropping** is adopted when other climatic factors are not limiting. Farm enterprise like dairy, poultry etc also influence the type of cropping system. When the farm enterprises include dairy, cropping system should contain fodder crops as components. Change in cropping system take place with the developments of technology.

Crop production, therefore, presents such an enormous diversity owing to differences in latitude, altitude and variability of rainfall and **edaphic diversity.** Thus, it may not be possible to enumerate and describe every type of **cropping pattern** prevalent in the country. Some broad contours of farming, however, emerge. Most important element of farming in India is production of foodgrains. On this basis, the country may be divided broadly into five agricultural regions.

1. The **rice region** extending from the eastern part to include a very large part of the north-eastern and the south-eastern India, with another strip along the western coast.
2. The **wheat region,** occupying most of the northern, western and central India.
3. The **millet-sorghum region,** comprising Rajasthan, Madhya Pradesh and the Deccan Plateau in the centre of the Indian Peninsula.
4. The **temperate Himalayan region** of Kashmir, Himachal Pradesh and Uttar Pradesh and some adjoining areas. Here potatoes are as important as cereal crops (which are mainly maize and rice) and the tree-fruits form a large part of agricultural production.
5. The **plantation crops region** of Assam and the hills of southern India where good quality tea is produced. There is an important production of high-quality coffee in the hills of the western Peninsular India. Rubber is mostly grown in Kerala and parts of Karnataka and Tamil Nadu. There are some large estates, but most of the growers would come under the category of small holders. Sugarcane, which in many countries is a plantation crop, is almost entirely grown by small holders in India.

1.11.2 Evaluation of Cropping System and Pattern

Various types of cropping system are practiced in a farm/region. They are to be properly evaluated to find out their stability and relative advantage. Such a comparison may be made with reference to land use efficiency, biological potential, economic viability, etc. Some of the important indices to evaluate the cropping system are discussed here.

Land Use Efficiency

Multiple cropping index (MCI) or multiple cropping: It was proposed by Dalrymple (1971). It is the ratio of total area cropped in a year to the total land area available for cultivation and expressed in percentage :

$$MCI = \frac{\sum_{(i=1)}^{n} a_i}{A} \times 100$$

where,

$i = 1, 2, 3, \ldots\ldots\ldots, n.$

n = Total number of crops,

a_i, = Area occupied by ith crop.

A = Total land area available for cultivation.

Cultivated land utilisation index (CLUI): Cultivated land utilisation index (Chuang, 1973) is calculated by summing the products of land area planted to each crop, multiplied by the actual duration of that crop divided by the total cultivated land area, times 365 days.

$$CLUI = \frac{\sum_{(i=1)}^{n} a_i d_i}{A \times 365}$$

where,
i = 1, 2, 3,........, *n*.
n = Total number of crops.
a_i = Area occupied by *i*th crop.
A = Total cultivated land area available for 365 days.

CLUI can be expressed as a fraction or percentage. This gives an idea about how the land area has been put into use. If the index is 1 (100%), it shows that the land has not been left fallow and more than 1, speaks about the specification of inter-cropping and relay cropping. Limitation of CLUI is its inability to consider the land temporarily available to the farmer for cultivation.

Crop intensity index (CII): Crop intensity index assesses farmer's actual land use in area and time relationship for each crop or group of crops compared to the total available land area and time, including land that is temporarily available for cultivation. It is cultivated by summing the product of area and duration of each crop divided by the product of farmer's total available cultivated land area and time period plus the sum of the temporarily available land area with the time of these land areas actually put into use. The basic concept of CLUI and CII are similar. However, the latter offers more flexibility, when combined with appropriate sampling procedures for determining and evaluating vegetable production and cropping pattern data.

$$CII = \frac{\sum_{(i=1)}^{n} a_i}{A} \times 100$$

where, i = 1, 2, 3, ..., N_c

N_c - Total number of crops grown by a farmer during the time period T,

a_i = Area occupied by *i*th crop,

t_i = Duration of *i*th crop (months that the crop z occupied an area a_i),

T = Time period under study (usually one year),

A_0 = Total cultivated land area available with the farmer for use during the entire time period, *T*,

M = Total number of fields temporarily available to the farmer for cropping during time period *T*,

j = 1, 2, 3,........ *M*,

Aj - Land area of *j*th field and *Tj* = Time period *Aj* is available.

CII = I means that area or land resources have been fully utilised and less than 1 indicates under utilisation of resources.

Biological Potential

Crop equivalent yield (CEY): Many types of crops/cultivars are included in a multiple cropping sequence. It is very difficult to compare the economic produce of one crop to another. To cite an example, yield of rice cannot be compared with potato yield. Similarly, the yield of crops grown for fodder purposes cannot be compared with the yield of grain .cereals or pulse crops and so on. In such situations, comparisons can be made based on economic returns (gross and net returns). Yields of protein and carbohydrate equivalents can also be calculated for valid comparison. Efforts have also been made to convert the yields of different crops into equivalent yield of any one crop, such as **wheat equivalent yield.** Equation for calculating wheat equivalent yield **(WEY)** is:

$$\text{WEY} = \sum_{(i=1)}^{n} Y_i e_i$$

where, Y_i = the economic yield of *i*th crop.

E_i = the wheat equivalent factor of *i*th crop, $e_i = \frac{P_i}{P_w}$

P_i= Price of unit weight of *i*th crop.

P_w = Price of a unit weight of wheat.

This type of comparison is valid when considering the gross returns. However, it does not indicate the net gain obtained from a cropping system. This will not give any explanation about the land use pattern of the cropping system.

Some common indices have been developed to measure and compare farmland use and production potential in multiple cropping. Land equivalent ratio (LER) is the most important one.

Land equivalent ratio (LER): This is the most frequently used efficiency indicator. LER can be defined as the relative area of sole crop that would be required to produce the equivalent yield achieved by inter-cropping.

$$LER = \sum_{(i=1)}^{n} \frac{Y_{ab}}{Y_{aa}} + \frac{Y_{ba}}{Y_{bb}}$$

where, Y_{ab} = Yield of crop *a* in inter-cropping.

Y_{ba} = Yield of crop *b* in inter-cropping.

Y_{aa} = Yield of crop *a* in pure stand.

Y_{bb} - Yield of crop *b* in pure stand.

LER of more than 1 indicates yield advantage, equal to 1 indicates no gain or no loss and less than 1 indicates yield loss. It can be used both for replacement and additive series of intercropping.

Economic Viability

Gross returns or gross profit: Total monetary returns of the economic produce, such as grain, tuber, bulb fruit, etc. and by-products *viz* straw, fodder, fuel etc obtained from the crops included in the system are calculated based on the local market prices. Total return is expressed in terms of unit area, usually one hectare.

The main draw back in this calculation is that the market price of the produce is higher than that actually obtained by the farmer. Generally, gross returns calculated is somewhat inflated compared to the actual receipt obtained by the farmer.

Net returns or net profit: This is worked out by subtracting the total cost of cultivation from the gross returns. This value gives the actual profit obtained by the farmer. In this type of calculation, only the variable costs are considered. Fixed costs, such as rent for the land, land revenue, interest on capital etc are not included. For a realistic estimate, however, fixed costs should also be included.

Rupees per rupee invested: This is also called profit-cost ratio or input-output ratio.

$$\text{Returns per rupee invested} = \frac{\text{Gross returns}}{\text{Cost of cultivation}}$$

Per day return: This is called as income per day and can be obtained by dividing the net returns by number of cropping period (days).

$$\text{Per day returns} = \frac{\text{Net returns}}{\text{Cropping period (days)}}$$

This gives the efficiency of the cropping system in terms of monetary value. If the system is stretched over one year, the denominator can be replaced by 365 days and per day return for the whole year can be calculated.

1.12 Allied Enterprises and their Management

A large number of allied **enterprises** like cattle, buffalo, goat, sheep and poultry rearing are taken up along with the major component of crop production for increasing the farm productivity.

1.12.1 Diary Farming

Dairy farming is an important source of income to farmers. Besides producing milk and/or draft power, dairy animals are also good source of **farm yard manure,** which is good source of organic matter for improving **soil productivity.** Farm by-products in turn are gainfully utilised for feeding the animals.

Cattle Rearing

Cattle rearing in India are carried out under a variety of adverse climatic and environmental conditions. Cattle are broadly classified into four groups :

- *Draft breeds*: Bullocks of these breeds are good draft animals, but the cows are poor milkers. Examples include Nagore, Hallikar, Kangeyam, Mali.
- *Dairy breeds*: Cows are high milk yielders, but the bullocks are of poor draft quality. Examples are Sahiwal, Sindhi, Gir.
- *Dual purpse*: Cows are fairly good milkers and the bullocks are with good draft work capacity (Haryana, Ongole and Kankrej).
- *Exotic breeds*: The exotic breeds are high milk yielder (Jersey, Holstein-Friesian, Aryshire, Brown Swiss and Guernsey).

Buffaloes

Important dairy breeds of buffalo are Murrah, Nili Ravi (which has its home tract in Pakistan), Mehsana, Suti, Zafarabadi, Godavari and Bhadawari. Of these, Godavari has been evolved through crossing local buffaloes in coastal regions of Andhra Pradesh with Murrah.

Housing: Each cow requires 12 to 18 m^2 space and the buffaloes need '2 to 15 m^2. Dairy building should be located at an elevated place to facilitate easy drainage. The floor should be rough and have gradient of 2.5 cm for every 25 cm length.

Breeding and maintenance: Cows remain in milk for 9-10 months, the average calving interval being 16-18 months. A cow does not require more than 6-8 weeks of dry period. From the economic point of view, cow should ordinarily be bred during the second and third months after calving. The best time to serve a cow is during the last stage of heat. If artificially inseminated, it is better to inseminate for 3 days continuously to have better probability to conceive. Gestation period varies with individual cows and breeds and normally it is about 280 days.

In the case of buffaloes, lactation period last for 7-9 months. She buffalo comes to heat every 21-23 days. Gestation period is 310 days. Regular de-worming of calf is needed for buffalo maintenance.

Feeding: Cattle feed, generally, contains fibrous, coarse, low nutrient straw material called **roughage** and **concentrates** as well as green fodder round the calendar year to harvest potential yield.

Roughages: Dairy cattle are efficient user of the roughages and convert large quantities of relatively inexpensive roughage into milk. Roughages are basic need for cattle and include legumes, non-legume hays, straw and silage of legume and grasses.

Concentrates: Grains and by-products of grains and oilseeds constitute the concentrates. They are extensively used in dairy cattle ration. These include cereals (maize, sorghum, oats, barley), cotton seeds, industrial wastes (bran or wheat, rice, gram husk) and cakes of oilseeds (groundnut, sesame, rapeseed, soybean, linseed).

Vitamins and mineral mixtures: It is advisable to feed a supplement containing vitamins A and D. Mineral mixtures containing salt, Ca, Mg and P should also be provided in the ration. Ration per animal per day includes concentrate at 1 kg for 2 liters of milk yield, green fodder (20-30 kg), straw (5-7 kg) and water (32 litres).

1.12.2 Goat and Sheep Rearing

The system of sheep and goat rearing in India is different from that adopted in the developed countries. In general, smaller units are mostly maintained as against large scale in fenced areas in the developed countries.

Goat rearing

In India, goat rearing is associated with different system such as crop or animal-based, pastoral or sedentary, single animal or mixed herd, small or large scale. Goat is mainly reared for meat, milk, hide and skin. Goat meat is the preferred meat in the country. A goat on hoof (live goat) fetches a better price than a sheep on hoof.

Housing: Goats can be maintained under stall-fed conditions. Goats do not thrive on marshy or swampy ground. Goats are to be provided with a dry, comfortable, safe and secure place, free from worms and affording protection from excessive heat and inclement weather. Kitts are kept under large inverted baskets until they are old enough to run along with their mothers. Males and females are, generally, kept together. The space requirement for a goat is 4.5 to 5.4 m^2.

Breeding and maintenance: Goat matures in about 6-7 months. Breeding is allowed for buck at one year and doe after 10 months of age. Gestation period is 145-155 days. It gives birth to 1-3 kitts per time. Number of evings is three per 2 years. The kitts can be weaned after 30-45 days. Mother can be allowed for mating 45-60 days after evings. When the young ones attain a body weight of about 25-30 kg in about nine months, they can be sold.

Feeding: Goats are essentially browsers and eat plants, which any other animals won't touch. Goats eat 4-5 times that of their body weight. They eat more of tree fodder and hence 40-50 per cent of green fodder should contain tree leaf fodder and the rest with other grass species. Goats should be fed with concentrates of maize, wheat, horsegram, groundnut cake, fish meal and wheat bran. Common salt and vitamin mixtures should also be added.

Sheep rearing

Sheep consume large quantities of roughage, converting a relatively cheap food into a good cash product. Housing need not be elaborate or expensive. However, to protect the flock from predatory animals, the height of the fencing should be raised to two metres.

Breeds of Indian sheep: There are three types of sheep in India, based on the geographical division of the country.

- *The temperate Himalayan region:* Gurez, Karanah, Bhakarwal, Gaddi.

- *Dry western region:* Lohi, Bikaneri, Marwari, Kutchi, Kathiawari.
- *Southern region:* Deccani, Nellore, Bellary, Mandya, Bandur.

Breeding and maintenance: One ram can be maintained for 40-50 ewes. Unlike other farm animals, ewes, in general, do not come in heat at regular intervals throughout the year but are seasonal in this respect. Duration of the heat period will range from 1-3 days and 75 per cent of the ewes remain in heat for 21-39 hours. Optimum time of service is towards the end of heat period. Average heat interval is 18 days during the breeding season. Gestation period will vary from 142-152 days with an average of 147 days. A normal ram is in full vigour for breeding from the age of 21/2-5 years. Sheep grow fully at two years of age, when the ewe is ready for breeding. Under average range conditions, ewes may be expected to produce about five crops of lambs.

Feeding: A sheep requires about 1-2 kg of leguminous hay per day depending on the age and its body weight. Proteins may be supplied through concentrates, such as groundnut cake, sesame cake or safflower cake when the pastures are poor in legumes or when scarcity conditions prevail. Normally, 110-225 g of cake is sufficient to maintain an average sheep in good condition. Feeding a mixture of common salt, ground limestone and sterilised bone meal in equal parts is required to alleviate deficiency of minerals in the feed.

1.12.3 Piggery

Pigs are maintained for the production of pork. They are fed with inedible feeds, forages; certain grain by-products obtained from mills, meat by-products, damaged feeds and garbage. Most of these feeds are either not edible or not very palatable to human beings. Pig grows fast and is a prolific breeder, farrowing 10 to 12 piglets at a time. It is capable of producing two litters per year under good management conditions. The carcass return is high at 65-70 per cent of the live weight.

Breeds: Imported breeds of Large White Yorkshire and Landrace are being used widely. Yorkshire is the most extensively used exotic breed in India. A good boar weighs 90 kg in about 5-6 months and is strong on feet and legs. Mother of the pig to be selected should have large litters of eight piglets or more.

Housing: Locally available materials can be used for housing. One pig requires about 2.7 m^2 with a wall of 1.2 m height. Eight boars can be kept in 2.7-4.5 m^2 area with 2.4-6.0 m^2 open space.

Feeding: Pigs are the most rapidly growing livestock and suffer more from nutritional deficiencies than the ruminants. Main ingredients of swine ration are cereals and millets and their by-products. Fiber content in swine ration should be very low (around 5-6%) for better feed utilisation efficiency. Mixed ration should also contain 0.5 per cent of added salt. Swine requires comparatively higher percentage of Ca and P than do cattle or sheep. On an average, the consumption of feed is 3.5 per cent of total weight. Feed allowance is calculated as 2.5- 3.0 kg 100^{-1} kg body weight + 0.25 kg feed per piglet with the lactating mothers.

Management: As a general rule, well-developed gilts weighing about 100 kg when 12- 14 months old, may be used for breeding. Gestation period is, on an average, 114 days. Litter size at birth may be 1-16 numbers with the body weight of 1-25 kg. Normal period between birth of piglets is 10-20 ihinutes. Time taken for the whole process of farrowing ranges from 1.5 to 4.0 hours. Sows are weaned after 40 days. Weaned sows come into heat in 3-10 days after weaning and may be allowed to breed. The boar-sow ratio should be 1:15. It is profitable to raise two litters from each sow each year.

1.12.4 Poultry

Poultry is one of the fastest growing food industries in the world. Poultry meat accounts for about 27 per cent of the total meat consumed. Average global consumption is 120 eggs per person per year and in India, it is only 32-33 eggs per capita year. As per the nutritional recommendation, per capita consumption is estimated at 180 eggs $year^{-1}$ and 9 kg meat $year^{-1}$.

Breeds: Majority of the stocks used for egg production are crosses involving the strains or inbred lines of white Leghorn. To a limited extent, other breeds like Rhode Island Red, California Grey and Australop are used.

Housing: Floor area of about 0.2 m^2 per adult bird is adequate for light breeds, such as white Leghorn. About 0.3-0.4 m^2 per bird is requited for heavy breeds. The house should have good ventilation and reasonably cool in summer and warm during winter. It should be located on well-drained ground, safe from flood waters.

Feed: About 60-70 per cent of the total expenditure on poultry farming is spent on the poultry feed. Hence, use of cheap and efficient ration will give maximum profit. Ration should be balanced containing carbohydrates, fats, minerals and vitamins. Some of the common feed stuffs used for making poultry ration in India are:

Cereals: Maize, barley, oats, wheat, pearlmillet, sorghum, rice-broken.

Cakes/meal: Oilcakes, maize-gluten-meal, fish meal, meat meal, blood meal.

Minerals/salt: Limestone, Oyster shell, salt, manganese.

From the day old to 4 weeks of age, birds are fed on starter ration and thereafter finisher ration, which contains more energy and 18-20 per cent protein. Feed may be given 2-3 times in a day. In addition to the foodstuffs, antibiotics and drugs may also be added to the poultry ration. Laying hens are provided with oyster shell or ground limestone. Riboflavin is particularly needed.

Maintenance: Chicks must be vaccinated against Ranikhet diseases with FI Strain vaccine within the first 6-7 days of age. One drop of vaccine may be administered in the eye and nostril. When chicks get the optimum body weight of 1.0-1.5 kg around six weeks, they can be marketed for broiler. Hens may be retained for one year for production (up to the age of about *Wi* years) since egg production would get reduced. One hen is capable of laying 180-230 eggs in a year starting from the six month.

1.12.5 Duck Rearing

Ducks are predominantly of indigenous type and reared for egg production on natural foraging. They have a production potential of about 130-140 eggs per bird per year. In places like marshy riverside, wetland and barren moors where chicken or any other type of stock do not flourish, duck rearing can be better alternative.

Breeds: Important Indian breeds are Sylhet Mete and Nageswari, which are mostly found in the eastern region of the country. Khaki Campbell and Indian Runner are the most popular breeds for egg laying. Indian Runner is the second best producer. White Pekin, Muscovy and Aylesbury are known for meat production.

Housing: In mild climate, it is possible to raise ducks without artificial shelter. A light fence of at least 1.2 m high enclosing the yard is enough to stop any predators. One nest of size 0.3 x 0.3 x 0.45 m to every 3 ducks is sufficient. A water channel of 0.5 m wide and 0.20 m deep is constructed at the far end on both sides parallel to the night shelter in the rearing or layer house.

Feeding: Ducks normally require lesser attention. They supplement their feed by foraging, eating fallen grains in harvested paddy fields, small fishes and other aquatic materials in lakes and ponds. However, for intensive rearing, pellet feeding may be given.

Maintenance: The general, management of ducks is similar to that of the chickens. The incubation period is 28 days. During the early part of the life, newly hatched ducklings require warm temperature under the natural or village conditions.

1.12.6 Apiculture

Apiculture is the science and culture of honeybees and their management. Apiculture is a subsidiary occupation and it is an additional source of income for farm families.

Species: The two bee species, of India are *Apis cerana indica* and *A mellifera. A cerana* is known as Indian bees, while *A mellifera* is known as European/westem bee.

Apis cerana: Serves commercial bee keeping in most parts of the country. It has instinctive behaviour of swarming and absconding. Its honey yield varies from 12 to 15 kg per hive per annum with foraging range between 0.8 and 1.0 km.

Apis mellifera: This species has achieved a great success in north-western states of India. Average honey production from this species is between 30 and 40 kg per hive per annum with foraging range extending up to 2-3 km.

Management: Flowers of large number of plants species are visited by honeybees for nectar and pollen. Most important sources of nectar and pollen are maize, mustard, sunflower and palm, litchi, pongamia, coconut, sesamum, etc. The beginner should start with 2 and not more than 5 colonies. The hive consists of bottom-board, brood chamber, brood chamber frames, super chamber, super chamber frames, top cover, inner cover, and entrance rod. The most suitable time for commencing bee keeping in a locality is the arrival of the swarming season. Swarming is a natural tendency of bees to divide their colonies under conditions that are generally favourable for the survival of both parent colony and the swarm. This occurs during the late spring or early summer.

Honey collection: Honey is a sweet viscous fluid produced by honeybees mainly from the nectar of the flowers. The nectar collected by bees is processed and placed in comb cells for ripening. During the ripening, sucrose is converted into glucose and fructose by an enzyme called invertase, which is added to it by the bees. Honey is an excellent energy food with an average of about 3500 calories per kg. It is directly absorbed into the human blood stream, requiring no digestion.

1.12.7 Fishery

With the traditional management, farmers obtain hardly 300-400 kg of wild and culture fish per ha annually. However, composite-fish culture with the stocking density of 5000-7500 fingerlings ha^{-1} and supplementary feeding can boost the total biomass production.

Pond: Depth of the pond should be 1.5-2.0 m. Clay soils have higher water retention capacity & hence, are best suited for fish rearing. Pond-water should have appropriate proportion of nutrients, phosphate (0.2-0.4 ppm), nitrate (0.06-0.1 ppm) and dissolved oxygen (5.0-7.0 ppm). Water should be slightly alkaline (pH 7.5-8.5). If the pH is less than 6.5, it can be adjusted with the addition of lime at regular interval of 2-3 days. Higher pH (>8.5) can be reduced with the addition of gypsum. Application of fresh dung may also reduce high pH in the water. Soil of the pond should be tested for N and P content. If the nutrient content is less, nitrogenous fertilisers like ammonium sulphate and urea and phosphatic fertiliser like superphosphate can be added. Organic manures such as FYM and poultry droppings may also be applied to promote the growth of phyto and zooplanktons. Species of fish are:

1. Catla (*Catla catla)* is the fast growing fish. It consumes lot of vegetation ajid decomposing higher plants. It is mainly a surface and column feeder.
2. Rohu (*Labeo rohita)* is a column feeder and feeds on growing fish. It consumes lot of vegetation and decomposing higher plants. It is mainly column and surface feeder.
3. Calbasu (*Labeo calbasu)* is a bottom feeder on detritus. Mrigal (*Cirrhinus mrigala)* is also a bottom feeder, taking detritus to large extent, diatoms, filamentous and other algae and higher plants. Common carp (*Cyprinus carpio*) is a bottom feeder and omnivorous.
4. Silver carp (*Hypophthalmichlthys molitrix*) is mainly a surface and phytoplankton- feeder and also feeds on micro-plants.

5. Grass carp (*Ctenopharyngodon idella)* is a specialised feeder on aquatic plants, cut- grass and other vegetable matter. It is also a fast growing exotic fish.
6. In composite fish culture, phytophagous fish (catla, rohu and mrigal) can be combined with omnivorous (common carp), plankton-feed (silver carp) and mud-eaters (mrigal and calbasu) in a composite fish culture system.

Management: For higher productivity, fish are to be provided with supplementary feeding with rice bran and oilseed cakes. This will enable faster growth and better yield. Each variety of carps could be stocked to 500 fingerlings with the total 5000-8000 ha^{-1}. This stocking density will enable to get a maximum yield of 2000 to 5000 kg ha ^{1} of fish annually under good management practices.

1.12.8 Sericulture

India occupies second position among silk producing countries in the world, next to China. The total area under mulberry is around 200 thousand ha in the country. More than 98 per cent of mulberry-silk is produced from five traditional sericultural states (Karnataka, Andhra Pradesh, West Bengal, Tamil Nadu and Jammu and Kashmir).

Moriculture: Cultivation of mulberry plants is called as **moriculture.** There are about 20 species of mulberry, of which four are commonly cultivated. They are *Morns alba, M indica, M serrata* and *M latifolia.* The crop can yield well for 12 years, after which they are pulled out and fresh planting is done. Yield of mulberry leaves is 30-40 t ha^{-1} year^{-1}.

Silk worm rearing: There are four types of silk worm: (1) Mulberry silk worm- *Bombyx mori,* (2) Eri silk worm-*Philosamia ricini,* (3) Tassar silk worm - *Antheraea mylitta* and (4) Muga silk *worm-Antheraea assami.*

Rearing and maintenance: Fertilised moth is covered with an inverted funnel or cellule and eggs are allowed to be laid over a cardboard. Parasites may be removed by brushing the egg masses with a fine brush. In a bamboo tray, rice husk is spread. Tender chopper mulberry leaves are added to the tray. The hatched out larvae are transferred to the leaves. It is important to change the leaves every 2-3 hours during the first 2-3 days.

The cocoon is constructed with a single reelable thread of silk. If the moths are allowed to emerge from the cocoons, the silk thread is cut into pieces. Hence, the pupa are killed 2- 3 days before the emergence of moth and processed. The cocoons required for further rearing are kept separately and moths are allowed to emerge from them.

1.12.9 Mushroom Cultivation

Mushroom is an edible fungus with great diversity in shape, size and colour. Essentially, mushroom is a vegetable that is cultivated in protected farms in a highly sanitised atmosphere. Just like other vegetables, mushroom contains 90 per cent moisture with high in quality protein. Mushrooms are fairly good source of vitamin C and B complex.

Species: There are three types of mushrooms popularly cultivated in India: (1) Oyster mushroom-*Pleurotous* sp, (2) Paddy straw *mushwom-Volvariella volvacea* and (3) White botton mushroom-*Agaricus bisporus.*

Oyster mushroom: Take fresh paddy straw and cut into small pieces of 3-5 cm length. Soak them in water for 4-6 hours and then boil for half an hour. Drain the water and dry the straw in shade till it is neither too dry nor wet. Take polythene bags of 60 x 30 cm size and make two holes of one cm diameter in the centre of the bag such that they face opposite sides. Tie the bottom of the bag with a thread to make a flat bottom. Fill the bag with paddy straw to 10 cm height and inoculate with the spawn. Likewise, prepare 4-5 layers of straw and spawn alternatively. Last layer ends up in straw of 10 cm height. Keep this in a spawn running room maintained at a temperature of about 22-28°C and with RH 85-90 per cent.

After 15-20 days when the spawn running is completed, cut open the polythene bag and take it to cropping room and allow it to grow for 7 days and harvest the mushroom. Mushroom yield is around 0.5-1.0 kg bag^{-1}.

Paddy straw mushroom: Cut the straw into long pieces of 60-90 cm and soak in water for 12 hours and sterilise 15 minutes. Arrange the straw in bundles. Lay the moistened straw bundles on the slightly raised concrete floor or on wooden platform in layers of four bundles width. Spawn or seed the beds simultaneously in each layer either by broadcasting or placing the grain spawn at different spots. Sprinkle grain dhal over each layer on the spawn. Don't spawn below the topmost layer. Maintain it at 30-35°C. Harvesting is ready after 25-30 days. Yield is around 1-1.5 kg bed^{-1}.

Botton mushroom: It requires a complex method of preparing compost, which is used as a substrate for mushroom production. Spawning is done by three methods: surface spawning, layer spawning and trough spawning. Fill the trays with compost and do spawning. After spawning, compost is pressed hard to make it compact. Trays are arranged in the cropping room in tiers and cover with newspaper sheet sprayed with 2 per cent formalin. Temperature of 20-25°C and RH of 90-95 per cent should be maintained. After spawn running is completed in 15-20 days and do casing. Pin heads appear within 10-15 days after casing. Cropping continues for 60-75 days. Mushrooms can be harvested at button stage. Yield ranges from 6-7 kg m^{-2}.

1.12.10 Agro-Forestry

Agro-forestry is a collective name for land-use system and technologies, in which woody perennials (trees, shrubs, palms, bamboos etc) are deliberately combined on the same land-management unit as agricultural crops and/or animals, either in some form of spatial arrangement or in a temporal sequence. In agro-forestry system, there are ecological and economical interactions among different components.

1. Agro-forestry normally involves two or more species of plants (or plants and animals) at least one of which is woody perennials.
2. An agro-forestry system always has two or more outputs.
3. The cycle of an agro-forestry system is always more than one year.
4. Even the simplest agro-forestry system is structurally, functionally and socio-economically more complex than a monocropping system.

Agro-forestry is important for meeting fodder, fuelwood and small timber of farmers, conserving soil and water, maintenance of soil fertility, controlling salinity and water-logging, positive environment impact and alternate land use for marginal and degraded lands. Selection of proper land-use system conserves biophysical resources of non-arable land besides providing day-to-day needs of farmer and livestock within the farming system. Commonly followed agro-forestry systems in India are:

1. Agri-silviculture (crops + trees), which is popularly known as farm forestry.
2. Agri-horticulture (crops + fruit trees).
3. Silvi-pasture (trees + pasture + animals).

4. Agri-horti-silviculture (crops + fruit trees + MPTs + pasture).
5. Horti-silvi-pasture (fruit trees + MPTs + pasture).
6. Agri-silvi-pasture (crops + trees + pasture).
7. Homestead agro-forestry (multiple combinations of various components).
8. Silvi-apiculture (trees + honey bees).
9. Agri-pisci-silviculture (crops + fish + MPTs).
10. Pisci-silviculture (fish + MPTs), etc.

Agri-silviculture: This system emphasises rising of trees and cultivation of field crops and/or fodder crops in the available space between the trees. In arid and semi-arid regions, hardy trees like *Prosopis cineraria* (khejri), *Eucalyptus sp, Acacia tortilis, Hardwickia binata* (anjan), *Azadirachta indica* (neem), *Ailanthes excelsa, Ziziphus jujuba* etc could be grown along with dryland crops, such as pulses (pigeonpea, blackgram), millets (pearlmillet, fingermillet, sorghum) etc. This is practiced mostly on arable lands, wherein multipurpose trees used for fuel and fodder can be grown with crops in the fields as alley farming. The hedges follow contour and compromise trees and shrubs like *Leucaena* or pigeonpea. Leguminous perennials are more suitable due to fixation of nitrogen.

Agri-horti-silviculture: In this system, fruit trees are grown along with crops and multipurpose trees (MPTs). Under rainfed situation, hardy fruit trees like ber, pomegranate, guava could be grown along with dryland crops like pigeonpea, *til,* mothbean, mustard, etc. Grafted ber (gola, seb, mundiya, banarasi kasak) may be planted at 6 x 6 m with 2 plants of subabul in between. Under partial irrigation, guava, pomegranate, lemon, etc. have been successfully grown at 6 x 5 m along with crop like wheat, groundnut and subabul (200 plants ha^{-1}) for quick leaf fodder and fuelwood production. For protection of fruit crops from desiccating hot summer and cold, winter planting of subabul/sesbania at every 2 m apart as windbreaks. Alternate plants of subabul/sesbania could be harvested for quick fodder and fuelwood production every 3rd year. Relative grain yield varies from 70 to 85 per cent even in 3rd and 4th year.

Silvi-pastoral system: In this system, improved pasture species are introduced with tree species. In this system, grasses or grass-legume mixture is grown along with the woody perennial simultaneously on the same unit of land. In the marginal, sub-marginal and other degraded lands, silvi-pastoral system has been found to be most economic agro-forestry system especially in arid and semi-arid regions. It involves lopping trees and grazing understory grasses and bushes in forests or plantations. It helps in reduction of the cost of concentrated feed to animal during lean period. A number of fodder trees like *Leucaena latisiliqua, Bauhinia variegata, Albizzia labbek, Albizzia amara, Moringa olerifera, Sesbania sesban, S grandiflora, Hardwickia binata* are identified for different regions of the country for silvi- pastoral system. Trees provide fuel and timber in the extreme dry season and lean periods, animal graze on pastures and feed on the leaves of nutritious trees and shrubs. Multi-layered vegetation covers are very effective in controlling runoff and soil loss from erosion prone areas.

Horti-pastoral system: It involves integration of fruit trees with pasture. In the degraded arid and semi-arid rangeland regimes, number of overgrazed plants of *Ziziphus nummularis* are found which could be successfully budded with improved variety of ber (gola, seb, umranb, banaras, kaska) besides planting MPTs like anjan, subabul, khejri along grasses and legumes like *Cenchrus, Lasiurus, Chrysopogon, Stylosanthes, Sirato,* etc.

Agri-silvi-pasture: It is a combination of agri-silviculture and silvi-pastoral system. In arid degraded lands of Rajasthan, Gujarat and Haryana, often dryland crops (pearlmillet, moth, *urad, til,* etc.) are grown in strips along with grass strips to avoid shifting sand reaching cropped area. MPTs could be introduced both in the pasture strips as well as in the crop strips, which besides protecting the crops from desiccating hot and cold wind would also provide leaf fodder, timber etc besides pasture when there is crop failure. Woody plants could be *Acacia Senegal,* ber, *anjan,* neem, etc. Grasses could be *Cenchrus, Lasiurus* and legume *Stylo sp.*

Pastoral silvicultural system: Integrated crop farming is practiced to meet the requirements of grasses and fodder for livestock. Pastoral silvicultural system is the practice in which grazing is the main component with scattered trees grown in the area. This practice is adopted in semi-arid regions of the country comprising the states of Andhra Pradesh, Tamil Nadu, Karnataka, Maharashtra and Madhya Pradesh.

Cultivators leave the fields fallow with existing trees and protect the same. *Dichanthium annulatum* is an important grass under this system. Important planted trees in the system are *Eucalyptus* hybrid, *Casuarina equisetifolia, Borassus flabellifa* and *Phoenix sylvestris.* Generally, trees are lopped for fuel and fodder. Custard apple, mango, *Zizyphus* and tamarind fruits are used for domestic consumption.

1.12.11 Bio-Gas

Bio-gas is a clean, unpolluted and cheap source of energy, which can be obtained by a simple mechanism and little investment. Gas is generated from the cow dung during anaerobic decomposition. Bio-gas generation is a complex biochemical process. The celluloitic material is broken down to methane and carbon dioxide by different groups of micri-organisms. It can be used for cooking purpose, burning lamps, running pumps, etc.

Selection of a model: Two main designs of biogas plants are the floating gas holder and fixed-dome types.

Float dome type: Different models are available in this category (KVIC vertical and horizontal, Pragati model, Ganesh model).

Fixed dome type.- Gas plant is dome shaped underground construction. Masonary gasholder is an integral part of the digester called dome. Gas produced in the digester is collected in dome at vertical pressure by displacement of slurry in inlet and outlet. Entire construction is made of bricks and cement. Models available in this category are Janata and Deen-Bandhu. Selection of a particular type depends on technical, climatological, geographical and economic factors prevailing in a given area.

Selection of size: Size of the biogas plant is decided by the number of family members and the availability of dung. One cubic metre capacity plant will need two to three animals and 25 kg of dung. Gas produced will meet the requirement of a family of 4-6 members. It would suffice to have a 2 m^3 plant to cater to the needs of a family of 6-10 members.

Site selection and management: The site should be close to the kitchen or the place of use. Plant should get clear sunshine during most part of the day. Generation of dung has a direct bearing on the quantity of gas generated. Amount of gas production is considerably higher in summer followed by rainy and winter seasons. Gas production would be maximum at a temperature between 30 and 35°C. If the ambient temperature falls below 10°C, gas production is reduced drastically.

Bio-gas slurry: Slurry is obtained after the production of bio-gas. It is enriched manure. Another positive aspect of this manure is that even after weeks of exposure to the atmosphere, the slurry does not attract fleas and worms.

❑❑❑

6

Sustainable Agriculture

Paul Hawken (1993) advised: "Leave the world better than you found it, take no more than you need, try not to harm life or the environment, make amends if you do."

"The first rule of sustainability is to align with natural forces or at least not try to defy them'".

The word "sustain," from the Latin *sustinere* (*sus-*, from below and *tenere,* to hold), to keep in existence or maintain, implies long-term support or permanence. As it pertains to agriculture, sustainable describes farming systems that are "capable of maintaining their productivity and usefulness to society indefinitely. Such system... must be resource-conserving, socially supportive, commercially competitive and environmentally sound."

The concept of sustainable agriculture set out by the Technical Advisory Committee (TAC) of the Consultative Group on International Agricultural Research (CGIAR) states "sustainable agriculture is the successful management of resources for agriculture to satisfy the changing human needs, while maintaining or enhancing the quality of environment and conserving natural resources".

The United States Department of Agriculture defined **'sustainable agriculture'** as a management system for renewable natural resources that provides food, income and livelihood for present and future economic productivity and eco-system services of these resources.

Sustainable agriculture is a farm system that mimics as closely as possible the complexity of a healthy and natural ecosystem. It integrates three main goals:

1. Environmental health.
2. Economic efficiency.
3. Social-economic equity (community).

Environmental Health

Ecological sustainability is associated with genetic resource base and **bio-diversity.** Increase in diversity at genetics, species and **ecosystem** level forms the backbone of a sustainable ecology. Several possible solutions to the environmental problems created by capital and technology-intensive farming system have been proposed and research is currently in progress to evaluate alternative system. Main focus lies on the reduction or elimination of agro-chemical inputs through changes in management to assure adequate plant nutrition, plant protection through nutrient sources and **integrated pest management** and efficient use of energy and water.

Economic Efficiency

In an economic contest, the farm must generate revenue, not always in terms of cash, but in terms of kind also. Farmers make decisions in risky, continuously changing environment and the consequences of their decisions are, generally, not known when the decisions are made. Variability of yields and prices are major sources of risk in agriculture. Changes in technology, legal and social concerns and the human factor itself also contribute to risky environment for the farmers. Excessive dependency on external inputs maximises the risk of profit reduction. In order to stabilise the farm economics, the farm should be made self sufficient, there should be parity in income distribution and finally the resilience of agricultural system should be increased.

Community

An ideal farming system will surely meet nutritional and livelihood security of the farmers and sustainability decreases the market dependency. It should have equitable access to 'common property resources'. A farm can be said to be socially just only when it satisfies all the following options:

- Whether the farm is getting fare share of price.
- Whether it is sufficient to maintain a family well-being.
- Whether he get enough remuneration to purchase his food.
- Whether the minimum wage requirement is attained.

The Three Legs of Sustainability

Imagine a 3-legged stool. What happens if one of the legs breaks or one leg is missing entirely? The whole stool falls over. The 3-legged stool has become a metaphor for the need to consider the economic, environmental and social impacts of agriculture (or any of our actions). If our agricultural system has unacceptable impacts in any one of these spheres, it can't support producers and contribute to the community over the long term.

In order to be sustainable, three areas must be addressed by our agriculture, food and natural resource system. These three areas are **economics, environment** and **community.** A sustainable agriculture must provide a fair and reasonably secure living for farm families. It should minimise harm to the natural environment. It should maintain basic natural resources, such as healthy soil, clean water and clean air. And it should support viable rural communities and fair treatment of all people involved in the food system, from farm workers to consumers.

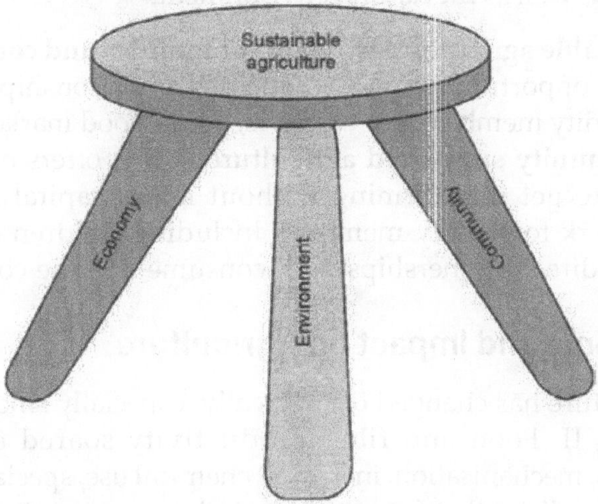

Fig. 2.1 : Three legs of sustainablility.

Sustainable agriculture can also be defined as: "an integrated system of plant anjl animal production practices having a site-specific application that will, over the long term:

1. Satisfy human food and fiber needs.
2. Enhance environmental quality and the natural resource base upon which the agricultural economy depends.
3. Make the most efficient use of non-renewable resources and on-farm resources and integrate, where appropriate, natural biological cycles and controls.

4. Sustain the economic viability of farm operations.
5. Enhance the quality of life for farmers and society as a whole."

Let's take a look at each individual "leg" of the sustainability stool.

Sustainable agriculture is "economically sustainable." Agriculture should provide a secure living to farm families and others employed in food production and processing. An economically sustainable approach also provides access to good food for all people.

Sustainable agriculture is "environmentally sound." It preserves the quality of basic natural resources that the farms, businesses and the surrounding environment rely on, including soil, water and air. Agriculture affects natural resources. Cooperating with natural resource system instead of trying to overpower them can offer benefits to food production as well as the natural environment.

Sustainable agriculture is "good for families and communities." It promotes opportunities and co-operative relationships for family and community members. For example, a local food marketing system called **community supported agriculture (CSA)** offers opportunities for people to get into farming without major capital investment, provides work for family members, including children on the farm and creates direct partnerships with consumers in the community.

2.1 Problems and Impact on Agriculture

Agriculture has changed dramatically, especially since the end of World War II. Food and fiber productivity soared due to new technologies, mechanisation, increased chemical use, specialisation and government policies that favoured maximising production. Although, these changes have had many positive effects and reduced many risks in farming, there have also been significant costs. Prominent among these are topsoil depletion, groundwater contamination, decline of family farms, continued neglect of the living and working conditions for farm labourers, increasing costs of production and the disintegration of economic and social conditions in rural communities. **Problems of agriculture** can be listed as under:

1. Decline in agricultural growth rate.
2. Decline in factor productivity.
3. Static or decline in food production.
4. Increasing malnutrition.

5. Shrinkage in net cultivated area.
6. Increasing environmental pollution.
7. Depleting groundwater table.
8. Increasing cost of production.
9. Low farm income.
10. Increasing unemployment.

As a result, agricultural productivity has been witnessing stagnation in recent years. Besides, issues, such as competing demand for water in the context of changing demographics and its various end uses further aggravates the degree of risks in the agriculture sector. These have considerable implications for food and livelihood security and as agriculture production being risk prone, may lead to migration from rural to urban and sub-urban areas.

Indian agriculture now faces the challenge of ensuring food security amidst constraints, such as stagnating net sown area, deterioration of land quality, reduction in per capita land availability, etc. The challenges are:

- Need for more food: Around 310 M t of foodgrains in 2050.
- Stagnating net sown area: 140 ± 2 M ha since 1970.
- Land share under fallow: 38 per cent increase since 1951.
- Per capita land availability: From 0.91 ha in 1951 to 0.32 ha in 2001 and to 0.19 ha is projected by 2050.
- Per capita net sown area: 0.33 ha in 1951 to 0.14 ha in 2001.

Fostering rapid, sustainable and broad-based growth in agriculture is therefore, a key priority keeping in mind the overall socio-economic development trajectory of the country, especially in the light of existing vulnerabilities that relate to a shrinking land resource base, additional stresses arising from the non-agricultural sector and issues emerging due to changing climate. This necessitates a strategic approach with a renewed vision and redefined focus.

A growing movement has emerged during the past three decades to question the role of the agricultural establishment in promoting practices that contribute to these social problems. Today, this movement for sustainable agriculture is gamering increasing support and acceptance within mainstream agriculture. Not only does sustainable agriculture address many environmental and social concerns, but also it offers innovative and economically viable opportunities for growers, labourers, consumers, policy-makers and many others in the entire food system.

2.2 Indicators of Agricultural Sustainability

Indicators of sustainability help us to identify, quantify and evaluate the effects of agriculture. We will be able to see the consequences of our policies on progress towards sustainability. Our future programmes and initiatives, as well as those of others, will be more easily and better targeted.

This set of indicators provides a means of measuring the economic, social and environmental impacts of agriculture and to help assess the effectiveness of policies and the sustainability of the sector.

In order to be able to make a balanced assessment of agriculture's progress towards sustainability, social, environmental and economic factors must be examined. This calls for indicators that are cross-linked as well as representative of the issues most relevant to agriculture. The trends revealed by individual indicators need careful interpretation before conclusions can be drawn on the progress towards sustainability of the sector as a whole.

Indicators are able to show positive progress towards sustainability. For example, use of environmentally friendly farming system has been rising. And by combining indicators on agricultural productivity and energy use, we can see that energy efficiency has been rising. But other indicators show adverse movement. The numbers of certain species of farmland bird, for example, are in decline. Some of the indicators included in this set are not yet fully developed as they are based on existing data. In time, however, they should provide a better reflection of the trends towards a more sustainable agriculture sector.

Indicators are a composite set of attributes or measures that embody a particular aspect of agriculture. Indicators are quantified information, which help to explain how things are changing over time. Sustainability indicators look at economic, social and environmental information in an integrated manner. Many professionals agree that at least three criteria should guide the development of sustainability indicators:

Table 2.1: Indicators of key natural resources in rainfed cropping system.

S No	Indicators key	Management aspects
1.	Nutrient balance	Organic matter -- rate of change.Nitrogen cycling—especially when using grain legumes in rotation with cereals. Monitoring status of phosphorus, sulphur and potassium. Micronutrients.
2.	Erosion	Vegetation cover—includes trees as well as stubble.Soil surface cover—stubble retained (30% sufficient to prevent wind and water erosion). Stream bank.Sheet and gully erosion.
3.	Productivity, yield and quality	Water use efficiency—*i.e.*, actual versus potential (in some areas the potential is much less than the actual) (biomass/grain yield/net return), recharge (dryland salinity and nutrient leaching).Pasture composition—legume and perennial.Matched animal versus pasture production—appropriate enterprise selection/ capability. Maintenance of genetic base/ improvement.
4.	Soil structure	Infiltration. Permeability / water storage. Stability. Waterlogging. Compaction.
5.	pH	Change.Toxicity-deficiency. Indicator plants.
6.	Energy efficiency	Energy input vis-a-vis energy output of the whole agricultural system.
7.	Biological factors	Soil macro/micro flora and fauna, animal health, plant health (root growth and other) pests (animals and plants).
8.	Farm management skills	Understanding a good indicator would be the understanding of the farmers of their own technical system.
9.	Precipitation (rainfall)	Performance of rainfall in a year as % of normal and its coefficient of variation.Distribution of area based on rainfall amount (dry: 0-750 mm, medium: 750-1150 and 1150-2000 mm, assured: > 2000 mm). Categorisation of the amount of rainfall (excess: + 20% or more, normal: + 19% to -19%, deficient: 20% to -59%, scanty: -60% or less).Number of districts having mean annual rainfall of 750-1250 mm and moisture availability period for at least 150 days.Area affected due to drought (slight, moderate or calamitous).

1. Policy relevance – indicators should address the issues of primary concern to a country or district and receive the highest priority. In some cases, policy-makers may already share concern about an aspect of sustainability (land degradation) and be ready to use indicator information for addressing the issue.
2. Predictability – to allow a forward-looking perspective that can promote planning and decisions on issues before they become too severe. Anticipatory decision making is at least as important to sustainable agriculture as is recognition of existing problems.
3. Measurability – to allow planners and analysts the means to assess how the indicator was derived, either qualitatively or quantitatively and decide how it can best be applied in the planning and decision-making process.

Some of the indicators of key natural resources in rainfed cropping system are listed in Table 2.1.

Most widely accepted framework for sustainability is referred to as Pressure/State/ Response (PSR framework), as described in the Table 2.2 and in the following flow diagram (Fig 2.2).

Table 2.2: Example of PSR framework.

Issue	Pressure (driving force)	Pressure (driving force)	Response (mitigating action)
Soil erosion	Hillside fanning	Declining yield	Terracing, perennial cropping.
Water quality	Agro-industrial processing	Fish die-off	Water treatment, technology adjustment.
Condition of grassland	Condition of grassland	Soil erosion	Stock rotation, destocking revegetation.

Pressure refers to the driving forces that create environmental impacts. They could include hillside farming, agro-industrial processing, livestock grazing, deforestation, etc.

State refers to the condition(s) that prevail when a pressure exists. This could be for example declining yields, fish die-off or soil erosion, etc.

Response refers to the mitigation action(s) and levers that could be applied to reduce or eliminate the impacts.

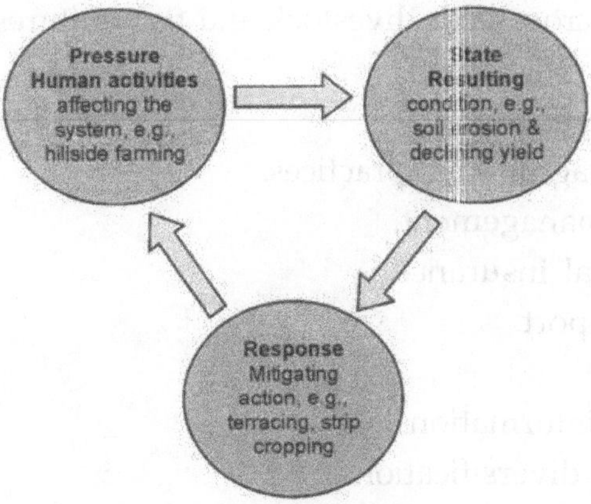

Fig. 2.2: The PSR framework for sustainability indicators (FAO research, extension and training).

2.3 Key Dimensions for Adaptation and Mitigation

Panel on Climate Change (PCC 2007) defines **mitigation** as the technological change and substitution that reduce resource inputs and emissions per unit of output. Although several social, economic and technological policies would produce an emission reduction, with respect to **climate change,** mitigation means implementing policies to reduce **GHG emissions** and enhance sinks.

Adaptation and mitigation measures needed for transforming Indian agriculture into a climate risk-resilient mode would have to focus on key dimensions in order to synergies strategies for programmatic interventions. These key dimensions (Fig 2.3) are the areas that need to be addressed due to the risks emanating from climate change. Each of these dimensions has then to be analysed in the context of four functional areas viz. research and development, technology and practices, infrastructure and capacity building for identifying adaptation and mitigation needs in a multi-dimensional and cross-sectoral matrix.

Collaborative programmatic interventions in these four areas focusing on the ten key dimensions would effectively address the adaptation and mitigation needs at different levels e.g. crop, farm and food system.

The key mitigation technologies in agriculture, as per National Mission for Sustainable Agriculture (NMSA 2010): Strategies for meeting the challenges of climate change, Department of Agriculture and Co-operation, Ministry of Agriculture New Delhi include:

1. Improved crop seeds, livestock and fish cultures.
2. Water use efficiency.
3. Pest management.
4. Improved agronomic practices.
5. Nutrient management.
6. Agricultural insurance.
7. Credit support.
8. Markets.
9. Access to information.
10. Livelihood diversification.

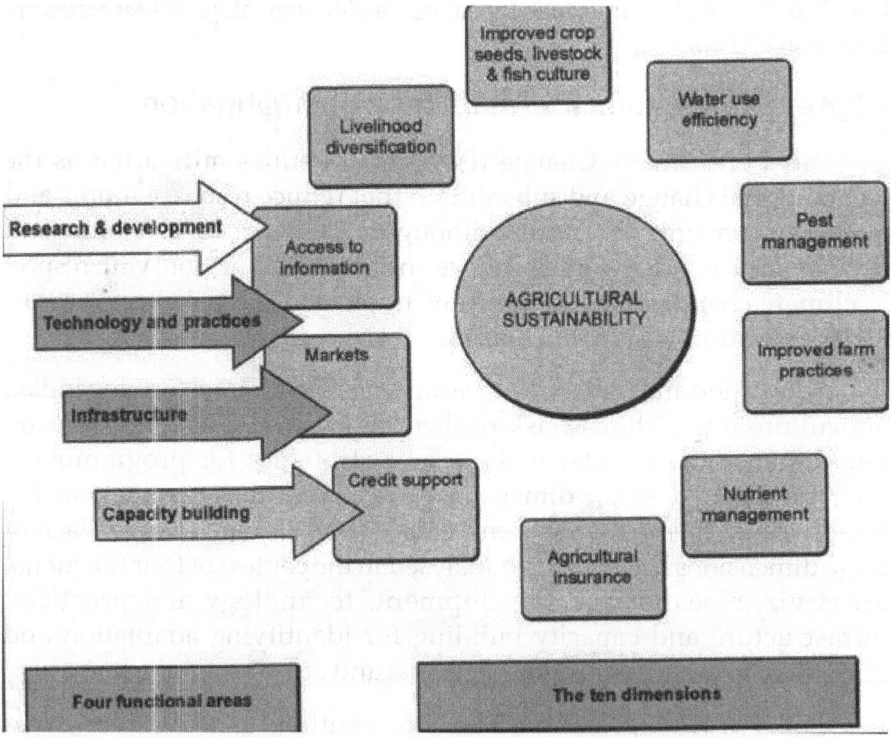

Fig. 2.3 : Key dimensions of agricultural sustainability.

2.3.1 Improved Crop Seeds, Livestock And Fish Cultures

- Promoting use of bio-technology.
- Research and promotion of C4 pathways in C3 plants.
- Conserving indigenous genetic resources.

- Public private partnerhsip (PPP) in R&D, management and dissemination of improved varieties.
- Conserving agricultural heritage.

Bio-technology has huge potential for combating vulnerabilities in crops, livestock and fisheries. Region specific vulnerability and impact assessment studies using crop growth simulation models are necessary to identify and formulate suitable interventions. Research and promotion of higher carbon (C4) pathways in low carbon (C3) plants and genetic manipulation of enzymes, such as RuBisco would help in increasing effectiveness of use of CO_2 and thus helping reduction in GHG emissions. Rich indigenous genetic resources in the crops, livestock and fisheries sectors should be conserved, catalogued and advantageously used while also conserving the agricultural heritage of the country. Relevant technologies should be used for rapid bulking of improved varieties of crops, livestock and fishes.

Strategies under this mission call for continuous efforts in improving genetic traits to sustain productivity, both in the short and long term, with focus on research and development of resilient genotypes. Simultaneously, reforms in policies, regulatory regimes and conformance to standards would also promote large-scale research and wider adoption of genetically improved varieties by both producers and consumers. International collaborations with Consultative Group on International Agricultural Research (CGIAR) institutions would be critical for fostering these activities.

2.3.2 Water Use Efficiency

- Promoting water use efficiency in irrigation.
- Research and development in the areas of energy efficient water system.
- Developing mechanisms for integrated management of rainwater, surface and groundwater.
- Policy instruments for PPP.
- Strengthen local institutions in managing water allocation and utilisation.

Two-thirds of the cultivated land in India is rainfed and suffers from water scarcity. Effective management of available water increasing **water use efficiency** and establishment of additional sustainable sources of water emerge as the primary issues that need to be addressed. Strategies under this dimension would focus on the application of a range of technologies coupled with demand and supply side management solutions to enhance water use efficiency for irrigation. While some technologies are available for direct application and can be implemented in the short term, there are other emerging areas like **recharging of aquifers, conjunctive use** of surface and groundwater, controlled extractions, etc. that would require collaboration and capacity building for technology absorption before being put into sustainable use.

2.3.3. Pest Management

- Efficient, safe and environmentally sound methods of pest management.
- Incentivising research, commercial production and marketing of **bio-pesticides.**
- Developing insect forecasting models.
- Decision and information support system for pest and disease surveillance.
- Institutional mechanism for quick response in case of disaster.

Pesticide consumption in India has increased over time and its injudicious use has created problems like development of resistant strains in insects and plant pathogens, resurgence of pest species, direct exposure to the applicator, destruction of parasites, predators and other beneficial organisms, accumulation of pesticide residues in agricultural commodities, water, air and soil, etc. **Pesticide residues** in feed and water affect livestock health due to direct and indirect exposure in the course of pest control measures.

Strategies suggested under this intervention have to primarily focus on establishment of decision and information support system for pest and disease surveillance, demonstration of best practices and quick response mechanism that are at par with the norms to deal with other disasters or natural calamities.

2.3.4 Improved Agronomic Practices

- Improved agronomic practices to reduce farm losses.
- Conservation and precision farming.

- Knowledge management.
- Soil conservation.
- Policy instruments for optimum land use.

Most effective way to address climate change is to adopt a sustainable development pathway by shifting to environmentally sustainable technologies and promotion and accelerated adaptation of energy efficient equipments, renewable energy and conservation of natural resources. Improved agronomic practices have the potential to help reduce farm level losses through improved soil treatment, increased water use efficiency, judicious use of chemicals, labour and energy and increased soil carbon storage. Targeted resource conserving technologies offer new opportunities for better livelihoods for the resource poor, small and marginal farmers. Improved agronomic mitigation practices include:

1. Improved crop and grazing land management to increase soil-carbon storage.
2. Restoration of cultivated peaty soils and degraded lands.
3. Improved rice cultivation techniques and livestock and manure management to reduce CH_4 emissions.
4. Improved nitrogen fertiliser application techniques to reduce N_20 emissions.
5. Dedicated energy crops to replace fossil fuel use.
6. Improved energy efficiency.
7. A large proportion of the mitigation potential of agriculture (excluding bio-energy) arises from soil carbon sequestration, which has strong synergies with sustainable agriculture and generally reduces vulnerability to climate change.
8. Considerable mitigation potential is also available from reductions in methane and nitrous oxide emissions in some agricultural system.
9. Biomass from agricultural residues and dedicated energy crops can be an important bio-energy feedstock, but current concerns with food prices make this a questionable alternative.

The strategies suggested under this intervention largely lean on support and popularisation of resource conservation technologies at the farm level and introduction of practices that would support mitigation efforts in time of extreme climatic events or disasters like prolonged dry spells, floods, etc. While in the short term, measures like changes in timing of farm operations and improved inputs to the farm can be employed, measures that include focus on conservation agriculture technologies and other best practices can be explored in the long run.

2.3.5. Nutrient Management

- Strengthening services for promoting production and use of **bio-fertilisers.**
- Developing nutritional strategies for managing heat stress in dairy animals.
- Strengthening capacity of existing soil testing labs.
- Quality standards and quality control system for raising confidence among users.

Plant nutrient management to increase soil nutrients and thus enhance crop productivity is a major technological challenge for ensuring food security and sustaining rural development. Plant nutrition management is essential to sustain and enhance crop productivity to meet the demand for food and raw materials and to maintain the quality of land and water resources. To ensure **soil health,** accurate inventorisation of soil resources is a prerequisite. Soil health can be improved through several site and soil-specific management options. Application of **integrated nutrient management** techniques has been found to increase **nutrient use efficiency** by integrating and balancing the nutrient dose in relation to nutrient status and crop requirements.

2.3.6 Agricultural Insurance

- Developing various models for risk assessment.
- Designing user-friendly decision support system to help assess risks and develop region specific contingency plans.
- Strengthening existing risk cover mechanism under NAIS and Weather Based Crop Insurance Scheme.
- Implementing region-specific contingency plans based on vulnerability and risk scenarios.

Agricultural insurance is an important mechanism by which risks to agricultural output and income can be addressed. Crop insurance incentive farmers should adopt innovative options by spreading the risks over space and time. It also stabilises farm incomes thereby enabling farmers to repay debts, which not only preserve the viability of formal financial institutions, but also saves huge government expenditures incurred in writing-off agricultural loans. Deficiencies in the existing framework of assessment of crop damage and prompt settlement of claims need to be addressed so that a *disaster mode of operational efficiency* is institutionalised. Research and development activities for developing new insurance products in the light of new risks emerging from **climate change** also need to be taken up as a medium to long term strategy. An effective design and efficient implementation mechanism is required to ensure timely benefits especially to the small and marginalised farmers.

2.3.7 Credit Support

- Developing new forms of credit assessment and risk management system.
- Promoting micro-finance.
- Developing mechanisms to enhance the flow of credit to critical infrastructure.
- Up-scaling the **kisan credit card scheme (KCCS).**
- Designing customised credit policies and programmes to mitigate risks.

Free, untied and timely credit support to farmers is essential for sustaining farm productivity, especially when it comes to small and marginal farmers. Easy and timely financial incentives and credit (and insurance) packages provided to farmers can help in adoption of improved management practices including resource conservation technologies, agro-diversification, post-harvest value addition processes etc which would contribute to reducing risks and enhancing farm incomes. This dimension emphasises efforts to augment the flow of credit to agriculture, alongside exploring new innovations in product design and methods of delivery, through better use of technology and related processes. Facilitating delivery through processors, input dealers, NGOs, self help groups (SHGs), etc. would help in providing access to credit to the resource poor farmers, especially the small and marginalised farmers, to help them to manage the additional risks from climate change in a sustainable manner.

2.3.8 Markets

- To formulate market-aligned research and development programmes.

- Improving supply chain efficiency.
- Creation of new market infrastructure.
- Supporting community partnerships in developing food and forage banks.
- Strengthening access to quality and timely inputs by farmers for mitigating risks.

Inadequate marketing infrastructure, presence of large number of intermediaries, lack of market information and intelligence and inadequate storage facilities results in huge post-harvest losses in the food supply chain. Some of the major initiatives that are to be taken up under this dimension include, reducing quantitative as well as qualitative losses across the supply chain, creating market aligned production system, strengthening climate resilient post-harvest management, storage and marketing and distribution system, strengthening timely access to farmers to quality inputs, strong farmer institution industry interface and encouraging food processing industries and greater exports.

2.3.9 Access to Information

- Minimising information asymmetry through ICT-based system.
- Public private partnership to develop technology based solution for providing farmers with information on price discovery, commodity arrivals, *mandi* prices, etc.
- Building an information and communication technology (ICT) enabled knowledge management network.
- To create, manage and develop national resource portal.

Effective communication approaches are critical to help farmers adapt to **climate change** as weather becomes more erratic and less predictable. Fresh strategies for management of information may be required to sustain production levels. This dimension is cross cutting in nature, having implications at all levels in the agricultural production system as well as for all the other key dimensions. At the crop level, the focus needs to be on up-scaling the efforts to link the public and private partners with the research institutions so that the laboratory results can get commercialised quickly. At the level of the farm, focus needs to be on enhancing awareness of farmers as well as the developmental agencies with the latest scientific research, market information and policy initiatives so that they are empowered to take informed decisions for maximising farm productivity. At a larger scale, at the food system level, technological and infrastructural research along with interventions required to enhance the adaptive capacity for ensuring food security in the wake of climate change must be investigated.

2.3.10 Livelihood Diversification

- Mitigating risks by supplementing income from off-farm activities.
- Crop diversification.
- Crop-livestock-fisheries farming system.

Livelihood diversification plays a major role in providing options of supplementing income from core agricultural activities through on-farm or off-farm activities, mitigating risks by providing additional support to agricultural income under conditions of climatic and non-climatic stresses, supporting farm-based investments for better productivity and through income generated by alternate livelihood options.

Strategies under this dimension would aim to promote diversification of agriculture into other high-value crops and horticulture, research, development and extension of crop-livestock farming system, increasing focus and development of approaches like sericulture, agro-forestry, crop-fish farming, etc.

2.4 Sustainable Agriculture : Related Terms

The following terms are commonly associated with sustainable agricultural system. None are synonymous with sustainable agriculture. Each relates to the concept in a different way. Most come under the umbrella of sustainable agriculture. A few, including bio-technology and precision farming, provoke much disagreement as to their appropriate use in a sustainable system. Some of the terms described are conceptual in nature, while others are strictly methodological. Many are combinations of both approaches. Several of these concepts and practices have very literal meanings that have been coloured by their historic use and practitioners' experiences.

Agro-ecology: Agro-ecology incorporates ideas about a more environmentally and socially sensitive approach to agriculture, one that focuses not only on production, but also on the ecological sustainability of the productive system.

Alternative farming/altentative agriculture: These are essentially synonymous terms encompassing a vast array of practices and enterprises, all of which are considered different from prevailing or conventional agricultural activities. They include :

- Non-traditional crops, livestock and other farm products.
- Service, recreation, tourism, food processing, forest/woodlot and other enterprises based on farm and natural resources (ancillary enterprises).
- Unconventional production system such as **organic farming** or **aquaculture** or direct marketing and other entrepreneurial marketing strategies."

Best management practices (BMPs): "BMPs were developed and implemented as a requirement of the 1977 amendments to the Clean Water Act. BMPs are established **soil conservation** practices that also provide water quality benefits. They include such practices as **cover crops, green manure crops** and **stripcropping** to control erosion along with **soil testing** and timing of chemical applications (similar to IPM) to prevent the loss of nutrients and pesticides. District soil conservation agents use BMPs in helping individual farmers develop conservation plans for their farms."

Bio-diversity: "At its simplest level, bio-diversity is the sum total of all the plants, animals, fungi and micro-organisms in the world or in a particular area; all of their individual variation and all the interactions between them."

Agro-bio-diversity: It is a fundamental feature of farming system around the world. It encompasses many types of biological resources tied to agriculture, including:

- **Genetic resources**—the essential living materials of plants and animals.
- Edible plants and crops, including traditional varieties, cultivars, hybrids and other genetic material developed by breeders and livestock (small and large, lineal breeds or thoroughbreds) and fresh water fish.
- Soil organisms vital to soil fertility, structure, quality and soil health.
- Naturally occurring insects, bacteria and fungi that control insect pests and diseases of domesticated plants and animals.

- **Agro-ecosystem** components and types (polycultural/ monocultural, small/large scale, rainfed/irrigated, etc.) indispensable for nutrient cycling, stability and productivity.
- Wild resources (species and elements) of natural habitats and landscapes that can provide services (for example, pest control and ecosystem stability) to agriculture.

"**Agro-bio-diversity** therefore includes not only a wide variety of species, but also the many ways in which farmers can exploit **biological diversity** to produce and manage crops, land, water, insects and biota."

Bio-dynamic agriculture/Bio-dynamic farming: Both a concept and a practice, bio-dynamics "owes its origin to the spiritual insights and perceptions of Dr Rudolf Steiner, an Austrian philosopher and scientist who lived at the turn of the century." Dr Steiner emphasised many of the forces within living nature, identifying many of these factors and describing specific practices and preparations that enable the farmer or gardener to work in concert with these parameters. "Central to the bio-dynamic method... are certain herbal preparations that guide the decomposition processes in manures and compost."

Biological farming/Ecological farming: Biological and ecological farming are terms commonly used in Europe and developing countries. Although, sometimes strictly defined, "biological farming is a system of crop production in which the producer tries to minimise the use of 'chemicals' for control of crop pests". Both **biological farming** and ecological farming are terms used in the broader sense, encompassing various and more specific practices and techniques of farming sustainability (organic, bio-dynamic, holistic, natural).

Bio-technology: Although farmers have been practicing bio-technology in the broadest sense (plant and animal breeding to achieve certain traits) for thousands of years, it is the recent breaking of the genetic code that has pushed this science into a new era altogether. Genetic engineering differs significantly from traditional bio-technological techniques in that DNA from different species can be combined to create completely new organisms.

Carbon sequestration: "Carbon sequestration is the process through which agricultural and forestry practices remove carbon dioxide (CO_2) from the atmosphere. The term 'sinks' is also used to describe agricultural and forestry lands that absorb CO_2, the most important global warming gas emitted by human activities. Agricultural and forestry practices can also release CO_2 and other greenhouse gases to the atmosphere. Sequestration activities can help prevent global climate change by enhancing carbon storage in trees and soils, preserving existing tree and soil carbon and by reducing emissions of CO_2, methane (CH_4) and nitrous oxide (N_2O)."

Carrying capacity: Carrying capacity is the theoretical equilibrium population size at which a particular population in a particular environment will stabilise when its supply of resources remains constant. It can also be considered to be the maximum sustainable population size; the maximum size that can be supported indefinitely into the future without degrading the environment for future generations.

Conservation buffer strips: Conservation buffer strips are areas or strips of land maintained in permanent vegetation, designed to intercept pollutants and erosion. Placed around fields, they can enhance wildlife habitat, improve water quality and enrich aesthetics on farmlands. Various types of buffers include contour buffer strips, filter strips, riparian forest buffers, field borders, windbreaks, shelterbelts, hedgerows, grassed waterways, and alley cropping.

Conservation tillage: Conservation tillage is a term that covers a broad range of soil tillage system that leaves residue cover on the soil surface, substantially reducing the effects of soil erosion from wind and water. These practices minimise nutrient loss and increase water storage capacity of the soil.

Ecological footprint (EF): Term has been introduced by William Rees in 1992. "It is a measure of how much land and water is needed to produce the resources we consume and to dispose of the waste we produce."

Carbon footprint: A representation of the effect human activities have on the climate in terms of the total amount of greenhouse gases produced (measured in units of carbon dioxide).

Environmental indicators: There are diverse interpretations as to what constitute environmental indicators and how they should be used. In any system, however, the "goal of environmental indicators is to communicate information about the environment and about human activities that affect it—in ways that highlight emerging problems and draw attention to the effectiveness of current policies... an indicator must reflect changes over a period of time keyed to the problem, it must be reliable and reproducible and whenever possible, it should be calibrated to the same terms as the policy goals or targets linked to it."

An agri-environmental indicator measures change either in the state of environmental resources used or affected by agriculture or in farming activities that affect the state of these resources. Examples of sustainable agriculture processes monitored by such indicators are soil quality, water quality, agro-ecosystem bio-diversity, climatic change, farm resource management and production efficiency.

Good agricultural practices (GAP): "Broadly defined, a GAP approach aims at applying available knowledge to addressing environmental, economic and social sustainability dimensions for on-farm production and post-production processes, resulting in safe and quality food and non-food agricultural products. Based on generic sustainability principles, it aims at supporting locally developed optimal practices for a given production system based on a desired outcome, taking into account market demands and farmers constraints and incentive to apply practices. However, the term "GAP" has different meanings and is used in a variety of contexts. For example, it is a recognised terminology used in international regulatory frameworks as well as in reference to private, voluntary and non-regulatory applications that are being developed and applied by governments, civil society organisations and the private sector."

Holistic management® (HM): "Holistic Management® originated when HMI's founder, Allan Savory, set out as a young wildlife biologist in his native Zimbabwe to solve the riddle of desertification. His efforts resulted in the development of the Holistic Management® Model-a proven, whole farm/whole system approach to resource management that incorporates financial planning, land planning, grazing planning and biological monitoring. The process increases soil health, reduces erosion, improves bio-diversity and enhances productivity by working with nature."

Integrated farming system (IFS)/integrated food and farming system (IFFS): Farming research and policy programs have begun to recognise that by viewing farms and the food production system as an integrated whole, more efficient use can be made of natural, economic and social resources. Included in this concept are the goals of finding and adopting "integrated and resource-efficient crop and livestock system that maintain productivity, that are profitable and that protect the environment and the personal health of farmers and their families," as well as "overcoming the barriers to adoption of more sustainable agricultural system such that these system can serve as a foundation upon which rural American communities will be revitalised."

Low input agriculture: Low input farming system "seek to optimise the management and use of internal production inputs (on-farm resources)... and to minimise the use of production inputs (off-farm resources), such as purchased fertilisers and pesticides, wherever and whenever feasible and practicable, to lower production costs, to avoid pollution of surface and groundwater, to reduce pesticide residues in food, to reduce a farmer's overall risk and to increase both short- and long-term farm profitability."

Natural fanning: Natural farming reflects the experiences and philosophy of Japanese farmer Masanobu Fukuoka. "His farming method involves no tillage, no fertiliser, no pesticides, no weeding, no pruning and remarkably little labor! He accomplishes all this (and high yields) by careful timing of his seeding and careful combinations of plants (polyculture).

Nature farming: Nature farming grew out of the philosophy and methodology of Japanese philosophise Mokicho Okada in the mid-1940s. "The theory of nature farming, as Okada expounded it, rests on a belief in the universal life-giving powers that the elements of fire, water and earth confer on the soil. The planet's soil, created over a span of eons, has acquired life-sustaining properties, in accordance with the principle of the indivisibility of the spiritual and the physical realms, which in mm provide the life- force that enables plants to grow. To utilise the inherent power of the soil is the underlying principle of nature farming."

Organic farming: The term 'organic farming' was first used by Lord Northboume in the book, *Look to the Land.* Lord Northboume, who embraced the teachings of Rudolph Steiner and bio-dynamic farming, had a "vision of the farm as a sustainable, ecologically stable, self-contained unit, biologically complete and balanced—a dynamic living oiganic whole. The term, thus, did not refer solely to the use of living materials (organic manures, etc.) in agriculture although obviously it included them, but with its emphasis on 'wholeness' is encompassed best by the definition 'of, pertaining to or characterised by systematic coordination of parts of the one whole.'

Permaculture: A contraction of "permanent agriculture," the word "permaculture" was coined by Australian Bill Mollison in the late 1970s. One of the many alternative agriculture system described as sustainable, permaculture is "unique in its emphasis on design; that is, the location of each element in a landscape and the evolution of landscape over time. The goal of permaculture is to produce an efficient, low-maintenance integration of plants, animals, people and structure... applied at the scale of a home garden, all the way through to a large farm."

Precision farming/agriculture: Precision agriculture is a "management strategy that employs detailed, site-specific information to precisely manage production inputs. This concept is sometimes called precision agriculture, prescription farming, site-specific management. The idea is to know the soil and crop characteristics unique to each part of the field and to optimise the production inputs within small portions of the field. The philosophy behind precision agriculture is that production inputs (seed, fertiliser, chemicals, etc.) should be applied only as needed and where needed for the most economic production."

Regenerative agriculture: Robert Rodale coined this term and it subsequently was expanded to "regenerative/sustainable agriculture" by the Rodale Institute and Rodale Research Centre. Two reasons given for the emphasis on "regenerative" are (1) "enhanced regeneration of renewable resources is essential to the achievement of a sustainable form of agriculture," and (2) "the concept of re-generation would be relevant to many economic sectors and social concerns."

Sustainable development: During the past 10 years, considerable interest in sustainability as applied to all areas of human activity has emerged worldwide. In response, President Clinton created the President's Council on Sustainable Development (PCSD) in April 1993. The council members adopted the Brundtland Commission definition, saying that sustainable development must ... "meet the needs of the present without compromising the ability of future generations to meet their own needs".

2.5 Conservation Agriculture

Tillage facilitates sowing a crop by loosening and aerating the soil and mixing residue from the previous harvest, soil organic matter and other nutrients, such as manure and chemical fertilisers evenly throughout the upper layers of the soil within the root zone of the coming crop. Tillage destroys weeds, dries the soil before seeding and creates a smooth, uniform soil surface that facilitates planting.

Soil **tillage** began **10-12** millennia ago in the Middle East, where early farmers used a digging stick to loosen the soil before planting seeds. The tool evolved into spade and then a triangular blade, its material evolving from wood to stone and finally metal. The first wooden plows were likely pulled by farmers themselves, with animals beginning to pull plows around 3,000 years ago in Mesopotamia.

Through most of history, **tillage** had more in common with current conservation agriculture than with modem ploughing practices. Cultivation is hard work, even with draft animals. A smallholder farmer plowing a field with animal traction typically has to walk 30- 40 km and the constant need to keep the animal and plow in line makes this no easy task. Not surprisingly, farmers chose to plow as little as was necessary to produce a good crop.

Then came Jethro Tull (1674-1741), the English agriculturalist who revolutionised horse- drawn farm machinery and ploughed the soil to release its nutrients, thinking this minimised the need to apply manure and the many weed seeds it contained. Ploughs became increasingly sophisticated through the 17th and 18th century, by the end of which German, Dutch and British improvements brought into being moldboards that turned the soil 135° and controlled weeds very efficiently.

European colonial powers took the modem plow to North and South America, Asia, Australia and Africa, where it became a primary tool for developing newly cultivated land. With the advent of tractors, operators no longer walked behind their plows but sat in front of them. Many farmers and agricultural experts started to believe that the more they tilled, the higher the yield they got. Powerful ploughs became icons of human conquest of nature.

Some perceptive farmers and experts began to discover that heavy ploughing, the same technique that brought bounteous harvests and wealth to Europe and other farming areas in cooler temperate climates, condemned farmlands in warmer climates to soil erosion and degradation. In the late 1940s, the introduction of plant growth regulators developed during World War II made possible the development of modem conservation tillage.

What is Conservation Agriculture?

Conservation agriculture (CA) is a concept for resource-saving agricultural crop production that strives to achieve acceptable profits together with high and sustained production levels while concurrently conserving the environment. The CA is based on enhancing natural biological processes above and below the ground. Interventions, such as mechanical soil tillage are reduced to an absolute minimum and the use of external inputs such as agro-chemicals and nutrients of mineral or organic origin are applied at an optimum level and in a way and quantity that does not interfere with or disrupt the biological processes. The CA is characterised by three principles which are linked to each other, namely:

1. Continuous minimum mechanical soil disturbance.
2. Permanent organic soil cover.
3. Diversified crop rotations in the case of annual crops or plant associations in case of perennial crops.

Conventional "arable" agriculture is, normally, based on soil tillage as the main operation. Most widely known tool for this operation is the plough, which has become a symbol of agriculture. Soil tillage has in the past been associated with increased fertility, which originated from the mineralisation of soil nutrients as a consequence of soil tillage. This process leads in the long-term to a reduction of soil organic matter. Soil organic matter -not only provides nutrients for the crop, but it is also, above all else, a crucial element for the stabilisation of soil structure. Therefore, most soils degrade under prolonged intensive arable agriculture. This structural degradation of the soils results in the formation of crusts and compaction and leads in the end to **soil erosion.** The process is dramatic under tropical climatic situations but can be noticed all over the world. Mechanisation of soil tillage, allowing higher working depths and speeds and the use of certain implements like ploughs, disk harrows and rotary cultivators have particularly detrimental effects on soil structure. Excessive tillage of agricultural soils may result in short term increases in fertility, but will degrade soils in the medium term. Structural degradation, loss of organic matter, erosion and falling bio-diversity are all to be expected.

Soil erosion resulting from soil tillage has forced us to look for alternatives and to reverse the process of **soil degradation.** The logical approach to this has been to reduce tillage. This led finally to movements promoting **conservation tillage** and especially **zero- tillage,** particularly in southern Brazil, North America, New Zealand and Australia. Over the last two decades, the technologies have been improved and adapted for nearly all farm sizes, soils, crop and climatic zones. Experience is still being gained with this new approach to agriculture and FAO has supported the process for many years.

Experience has shown that these techniques, summarised as **conservation agriculture** (CA) methods, are much more than just reducing the mechanical tillage. In a soil that is not tilled for many years, the crop residues remain on the soil surface and produce a layer of; mulch. This layer protects the soil from the physical impact of rain and wind but it also stabilises the soil moisture and temperature in the surface layers. Thus, this zone becomes a habitat for a number of organisms, from larger insects down to soil borne fungi and Tacteria. These organisms macerate the mulch, incorporate and mix it with the soil and decompose it so that it becomes humus and contributes to the physical stabilisation of the soil structure. At the same time, this soil organic matter provides a buffer function for water and nutrients. **Larger components of the soil fauna, such as earthworms,** provide a soil structuring effect producing very stable soil aggregates as well as uninterrupted macropores leading from the soil surface straight to the subsoil and allowing fast water infiltration in case of heavy rainfall events. Keeping the soil covered and planting through the mulch will protect the soil and improve the growing environment for the crop.

This process carried out by the edaphon, the living component of a soil, can be called **"biological tillage"**. However, biological tillage is not compatible with mechanical tillage and with increased mechanical tillage, the biological soil structuring processes will disappear. Certain operations such as mouldboard or disc ploughing have a stronger impact on soil life than others as for example chisel ploughs. Most tillage is, however, targeted at loosening the soil which inevitably increases its oxygen content leading in turn to the mineralisation of the soil organic matter. This inevitably leads to a reduction of soil organic matter which is the substrate for soil life.

Thus, agriculture with **reduced** or **zero mechanical tillage** is only possible, when soil organisms are taking over the task of tilling the soil. This, however, leads to other implications regarding the use of agro-chemicals. Synthetic pesticides and mineral fertiliser have to be used in a way that does not harm soil life. As the main objective of agriculture is the production of crops, changes in the pest and **weed management** become necessary with CA. Burning plant residues and **ploughing** the soil is mainly considered necessary for phytosanifary reasons: to control pests, diseases and weeds. In a system with reduced mechanical tillage based on mulch cover and **biological tillage,** alternatives have to be developed to control pests and weeds. Integrated pest management (IPM) becomes mandatory. One important element to achieve this is **crop rotation,** interrupting the infection chain between subsequent crops and making full use of the physical and chemical interactions between different plant species. Synthetic chemical pesticides, particularly **herbicides** are, in the first years, inevitable but have to be used with great care to reduce the negative impacts on soil life. To the extent that a new balance between the organisms of the farm-ecosystem, pests and beneficial organisms, crops and weeds, becomes established and the farmer learns to manage the **cropping system,** the use of synthetic pesticides and mineral fertiliser tends to decline to a level below that of the original **"conventional" farming system.**

Burning crop and weed residues destroy an important source of plant nutrients and soil improvement potential. The phytosanitary motives for burning and ploughing can better be achieved by integrated pest management practices and crop rotations.

2.5.1 Definition

The collective umbrella term commonly given to no-tillage, **direct-drilling, minimum- tillage** and/or **ridge-tillage,** to denote that the specific practice has a conservation goal of some nature. Usually, the retention of 30 per cent surface cover by residues characterises the lower limit of classification for **conservation-tillage,** but other conservation objectives for the practice include conservation of time, fuel, earthworms, soil-water, soil structure and nutrients. Thus, residue levels alone do not adequately describe all conservation tillage practices.

This has led to confusion among the agricultural scientists and more importantly, the farming community. To add to the confusion, the term **'conservation agriculture'** has recently been introduced and its goals defined by FAO.

Conservation agriculture (CA) aims to conserve, improve and make more efficient use of natural resources through integrated management of available soil, water and biological resources combined with external inputs. It contributes to **environmental conservation** as well as to enhanced and sustained agricultural production. It can also be referred to as resource efficient or **resource effective agriculture.**

According to FAO, CA is a set of practices that leave crop residues on the surface which increases water infiltration and reduces erosion. It is a practice used in **conventional agriculture** to reduce the effects of tillage on soil erosion. However, it still depends on **tillage** as the structure forming element in the soil. Never less, **conservation tillage** practices, such as **zero tillage** practices can be transition steps towards conservation agriculture.

Conservation agriculture (CA) is defined as minimal soil disturbance (no-till) and permanent soil cover (mulch) combined with rotations, is a recent agricultural management system that is gaining popularity in many parts of the world (FAO 2006).

Appropriate definition of CA appears to be "a sustainable agriculture production system comprising a set of farming practices adapted to the requirements of crops and local conditions of each region, whose farming and soil management techniques protect the soil from erosion and degradation, improve its quality and bio-diversity and contribute to the preservation of the natural resources, water and air, while optimising yields".

Distinguishing features of conventional and conservation agricultural system are given in Table 2.3.

Table 2.3 : Distinguishing features of conventional and conservation agricultural system

Conventional agriculture	Conservation agriculture
• Cultivating land, using science and technology to dominate nature	• Least interference with natural processes
• Excessive mechanical tillage and soil erosion	• No-till or drastically reduced tillage (biological tillage)
• High wind and soil erosion	• Low wind and soil erosion
• Residue burning or removal (bare surface)	• Surface retention of residues (permanently covered)
• Water infiltration is low	• Infiltration rate of water is high
• Use of *ex-situ* FYM/composts	• Use of *in-situ* organics/composts
• Green manuring (incorporated)	• Brown manuring/cover crops (surface retention)
• Kills established weeds, but stimulates more weed seeds to germinate	• Weeds are a problem in the early stages of adoption but decrease with time
• Free-wheeling of farm machinery, increased soil compaction	• Controlled traffic, compaction in tramline, no compaction in crop area
• Monocropping/culture, less efficient rotations	• Diversified and more efficient rotations
• Heavy reliance on manual labor, uncertainty of operations	• Mechanised operations, ensure timeliness of operations
• Poor adaptation to stresses, yield losses greater under stress conditions	• More resilience to stresses, yield losses are less under stress conditions
• Productivity gains in long-run are in declining order	• Productivity gains in long-run are in incremental order

2.5.2 Principles of Conservation Agriculture

The main **principles of conservation agriculture** are: Conservation agriculture system utilise soils for the production of crops with the aim of reducing excessive mixing of the soil and maintaining crop residues on the soil surface in order to minimise damage to the environment. By doing this CA will:

1. Provide and maintain an optimum environment of the root zone to maximum possible depth. Roots are able to function effectively and without restrictions to capture high amounts of plant nutrients and water.
2. Ensure that water enters the soil so that:
    - Plants never or for the shortest time possible, suffer water stress that will limit the expression of their potential growth.
    - Residual water passes down to groundwater and stream flow, not over the surface as runoff.
3. Favour beneficial biological activity in the soil in order to:
    - Maintain and rebuild soil architecture.
    - Compete with potential in-soil pathogens.
    - Contribute to soil organic matter and various grades of humus.
    - Contribute to capture, retention, chelation and slow release of plant nutrients.
4. Avoid physical or chemical damage to roots that disrupts their effective functioning.

The three principles of conservation agriculture include:

1. Direct planting of crop seeds.
2. Permanent soil cover, especially by crop residues and cover crops.
3. Crop rotation.

Direct Seeding or Planting

Direct seeding involves growing crops without mechanical seedbed preparation and with minimal soil disturbance since the harvest of the previous crop. The term **direct seeding** is understood in CA system as synonymous with no-till farming, zero-tillage, no-tillage, direct drilling, etc. Planting refers to the precise placing of large seeds (maize and beans); whereas seeding usually refers to a continuous flow of seed as in the case of small cereals (wheat and barley). Equipment penetrates the soil cover, opens a seeding slot and places the seed into that slot. Size of the seed slot and the associated movement of soil are to be kept at the absolute minimum possible. Ideally, the seed slot is completely covered by mulch again after seeding and no loose soil should be visible on the surface.

Land preparation for seeding or planting under no-tillage involves slashing or rolling the weeds, previous crop residues or cover crops or spraying herbicides for weed control and seeding directly through the mulch. Crop residues are retained either completely or to a suitable amount to guarantee the complete soil cover and fertiliser and amendments are either broadcast on the soil surface or applied during seeding.

Permanent Soil Cover

A permanent soil cover is important to:

- Protect the soil against the deleterious effects of exposure to rain and sun.
- Provide the micro and macro-organisms in the soil with a constant supply of "food".
- Alter the microclimate in the soil for optimal growth and development of soil organisms, including plant roots.

Cover crops need to be managed before planting the main crop. This can be done manually or with animal or tractor power. The important point is that the soil is always kept covered (FAO, 2006).

The effects of soil cover include:

1. Improved infiltration and retention of soil moisture resulting in less severe, less prolonged crop-water stress and increased availability of plant nutrients.
2. Source of food and habitat for diverse soil life: creation of channels for air and water, biological tillage and substrate for biological activity through the recycling of organic matter and plant nutrients.
3. Increased humus formation.
4. Reduction of impact of raindrops on soil surface resulting in reduced crusting and surface sealing.

5. Consequential reduction of runoff and erosion.
6. Soil re-generation is higher than soil degradation.
7. Mitigation of temperature variations on and in the soil.
8. Better conditions for the development of roots and seedling growth.

Means and practices include:

1. Use of appropriate/improved seeds for high yields as well as high residue production and good root development.
2. Integrated management and reduced competition with livestock or other uses (through increased forage and fodder crops in the rotation).
3. Use of various cover crops, especially multi-purpose crops, like nitrogen fixing, soil- porosity-re-storing, pest repellent, etc.
4. Optimisation of crop rotations in spatial, timing and economic terms.
5. Targeted use of herbicides for controlling cover crop and weed development.

Crop Rotations

Rotation of crops is not only necessary to offer a diverse "diet" to the soil micro-organisms, but as they root at different soil depths, they are capable of exploring different soil layers for nutrients. Nutrients that have been leached to deeper layers and that are no longer available for the commercial crop can be "recycled" by the crops in rotation. This way the rotation crops function as biological pumps.

Furthermore, a diversity of crops in rotation leads to a diverse soil flora and fauna, as the roots excrete different organic substances that attract different types of bacteria and fungi, which in turn, play an important role in the transformation of these substances into plant available nutrients. Crop rotation also has an important phytosanitary function as it prevents the carryover of crop-specific pests and diseases from one crop to the next via crop residues.

Effects of crop rotation include:

1. Higher diversity in plant production and thus in human and livestock nutrition.
2. Reduction and reduced risk of pest and weed infestations.

3. Greater distribution of channels or bio-pores created by diverse roots (various forms, sizes and depths).
4. Better distribution of water and nutrients through the soil profile.
5. Exploration for nutrients and water of diverse strata of the soil profile by roots of many different plant species resulting in a greater use of the available nutrients and water.
6. Increased nitrogen fixation through certain plant-soil biota symbionts and improved balance of N/P/K from both organic and mineral sources.

Means and practices include:

1. Design and implementation of crop rotations according to the various objectives: food and fodder production (grain, leaf, stalks), residue production, pest and weed control, nutrient uptake and biological subsurface mixing/cultivation, etc.
2. Use of appropriate/improved seeds for high yields as well as high residue production of above-ground and below-ground parts, given the soil and climate conditions.

2.5.3 Advantages of Conservation Agriculture

To be widely adopted, all new technology needs to have benefits and advantages that attract a broad group of farmers who understand the differences between what they are doing and what they need. In the case of conservation agriculture, advantages can be grouped as:

- Economic benefits that improve production efficiency.
- Agronomic benefits that improve soil productivity.
- **Environmental and social benefits** that protect the soil and make agriculture more sustainable.

Economic Benefits

Three major economic benefits can result from CA adoption:

1. Time saving and thus reduction in labour requirement.
2. Reduction of costs (fuel, machinery operating costs and maintenance, as well as a reduced labour cost).
3. Higher efficiency in the sense of more output for a lower input.

The positive impact of **conservation agriculture** on the distribution of labour during the production cycle and even more important, the reduction in labour requirement are the main reasons for farmers in Latin America to adopt conservation agriculture, especially for farmers who rely fully on family labour. Manual labour for soil preparation is back-breaking and unnecessary. Should the supply of labour be reduced, through sickness or migration, then the system can quickly become unsustainable.

Agronomic Benefits

Adopting conservation agriculture leads to improvement of soil productivity:

- Organic matter increase.
- In-soil water conservation.
- Improvement of soil structure and thus, rooting zone.

Constant addition of crop residues leads to an increase in the organic matter content of the soil. In the beginning this is limited to the top layer of the soil, but with time this will extend to deeper soil layers. Organic matter plays an important role in the soil: fertiliser use efficiency, water holding capacity, soil aggregation, rooting environment and nutrient retention, all depend on organic matter.

Environmental Benefits

1. Reduction in soil erosion and thus of road, dam and hydroelectric power plant maintenance costs.
2. Improvement of water quality.
3. Improvement of air quality.
4. Bio-diversity increase.
5. Carbon sequestration.

Residues on the soil surface reduce the splash-effect of the raindrops and once the energy of the raindrops has dissipated the drops proceed to the soil without any harmful effect. This results in higher infiltration and reduced runoff, leading to less erosion. Residues also form a physical barrier that reduces the speed of water and wind over the surface. Reduction of wind speed reduces evaporation of soil moisture.

Soil erosion is reduced close to the re-generation rate of the soil or even adding to the system due to the accumulation of organic matter.

Soil erosion fills surface water reservoirs with sediment, reducing water storage capacity. Sediment in surface water increases wear and tear in hydroelectric installations and pumping devices, which result in higher maintenance costs and necessitates earlier replacement.

More water infiltrates into the soil with conservation agriculture rather than running off the soil surface. Streams are then fed more by subsurface flow than by surface runoff. Thus, surface water is cleaner and more closely resembles groundwater in conservation agriculture than in areas where intensive tillage and accompanying erosion and runoff predominate. Greater infiltration should reduce flooding, by causing more water storage in soil and slow release to streams. Infiltration also recharges groundwater and thus, increasing well-supplies and revitalising dried up springs.

Sediment and dissolved organic matter in surface water must be removed from drinking water supplies. Less sediment loss and less soil particles in suspension, lead to a reduced cost for water treatment. Maintaining soil cover will reduce erosion with the consequent loss of soil fertility, soil compaction, and eventually, landscape change.

One aspect of **conventional agriculture** is its ability to change the landscape. Destruction of the vegetative cover affects the plants, animals and micro-organisms. However, most organisms are negatively affected and either they disappear completely or their numbers are drastically reduced. With the conservation of soil cover in conservation agriculture a habitat is created for a number of species that feed on pests, which in turn attracts more insects, birds and other animals. Rotation of crops and cover crops restrains the loss of genetic **bio-diversity,** which is favoured with **monocropping.**

System, based on high crop residue addition and no tillage, accumulate more carbon in the soil, compared to the loss into the atmosphere resulting from plough-based tillage. During the first years of implementing conservation agriculture, the organic matter content of the soil is increased through the decomposition of roots and the contribution of vegetative residues on the surface. This organic material is decomposed slowly and much of it is incorporated into the soil profile, thus the liberation of carbon to the atmosphere also occurs slowly. In the total balance, carbon is sequestered in the soil and turns the soil into a net sink of carbon. This could have profound consequences in the fight to reduce **green house gas** emissions into the atmosphere and thereby help to forestall the calamitous impacts of **global warming.**

2.5.4 Limitatins of Conservation Agriculture

Most important limitation in all areas where **conservation agriculture** is practiced is the initial lack of knowledge. There is no blueprint available for conservation agriculture, as all **agro-ecosystem** are different. Particularly important gap is the frequent dearth of information on locally adapted cover crops that produce high amounts of biomass under the prevailing conditions. The success or failure of conservation agriculture depends greatly on the flexibility and creativity of the practitioners and extension and research services of a region. Trial and error, both by official institutes and the farmers themselves, is often the only reliable source of information.

However, as conservation agriculture is gaining momentum rapidly in certain regions, there now exist networks of farmer organisations and groups of interested people who exchange information and experiences on cover crops, tools and equipment and other techniques used in conservation agriculture.

Initial nervousness about switching from plough-based farming to CA can be ameliorated by forming farmer groups to exchange ideas and gain knowledge from more experienced practitioners.

As conservation agriculture partly relies on the use of **herbicides,** at least during the initial stage of adoption, some people worry that adoption of conservation agriculture will increase herbicide use and that in turn will lead to increased contamination of water by herbicides. In fact, experience has shown that herbicide use tends to decline over time as the soil cover practices prevent weed emergence.

Reductions in leaching of pesticides under conservation agriculture might be caused by greater microbial activity degrading pesticides faster or to greater organic matter adsorbing the pesticides.

2.5.5 Constraints for Adopting Conservation Agriculture

A mental change of farmers, technicians, extensionists and researchers away from soil degrading tillage operations towards sustainable production system like no tillage is necessary to obtain changes in attitudes of farmers, however, noted that probably the most important factor in the adoption of CA is overcoming the bias or mindset about **tillage.** It is argued that convincing the farmers that successful cultivation is possible even with reduced tillage or without tillage is a major hurdle in promoting CA on a large scale. In many cases, it may be difficult to convince the farmers of potential benefits of CA beyond its potential to reduce production costs, mainly by tillage reductions. CA is now, considered a route to sustainable agriculture. Spread of conservation agriculture, therefore, will call for scientific research linked with development efforts. The following are a few important constraints which impede broad scale adoption of CA:

1. ***Lack of appropriate seeders especially for small and medium scale farmers:*** Although significant efforts have been made in developing and promoting machinery for seeding wheat in no till system, successful adoption will call for accelerated effort in developing, standardising and promoting quality machinery aimed at a range of crop and cropping sequences. These would include the development of permanent bed and furrow planting system and harvest operations to manage crop residues.
2. ***Wide spread use of crop residues for livestock feed and fuel:*** Especially under rainfed situations, farmers face a scarcity of crop residues due to less biomass production of different crops. There is competition between CA practice and livestock feeding for crop residue. This is a major constraint for promotion of CA under rainfed situations.
3. ***Burning of crop residues:*** For timely sowing of the next crop and without machinery for sowing under CA system, farmers prefer to sow the crop in time by burning the residue. This has become a common feature in the rice-wheat system in north India. This creates environmental problems for the region.
4. ***Lack of knowledge about the potential of CA to agriculture leaders, extension agents and farmers:*** This implies that the whole range of practices in conservation agriculture, including planting and harvesting, water and nutrient management, diseases and pest control, etc. need to be evolved, evaluated and matched in the context of new system.
5. ***Skilled and scientific manpower:*** Managing conservation agriculture system, will call for enhanced capacity of scientists to address problems from a system perspective and to be able to work in close partnerships with farmers and other stakeholders. Strengthened knowledge and information sharing mechanisms are needed.

2.5.6 Conservation Agriculture vs Conservation Tillage

Conservation agriculture is largely the product of the collective efforts of a number of previous agricultural movements, including no-till agriculture, agro-forestry, green manures/ cover crops, direct planting/seeding and conservation tillage among many others. Yet, CA is distinct from each of these so-called agricultural packages, even as it draws upon many of their core principles. This is because CA uses many of the available technologies in unison, resulting in something many believe to be much greater than the "sum of its parts." The following terms are often confused with conservation agriculture:

No-Till (NT)/ Zero-Till (ZT)

The NT and ZT are the technical components used in conservation agriculture that simply involve the absence of tillage/ploughing operations on the soil. Crops are planted directly into a seedbed not tilled after harvesting the previous crop. Not everyone utilising no-till technologies adopts other important components of CA. One major difference is that NT or ZT do not necessarily leave residue mulch. Some recent research data suggests this is vital, since without the residue mulch many of the benefits of CA are lost or decreased in value.

Conservation Tillage/ Minimum Tillage/ Reduced Tillage

These are tillage operations that leave at least 30 per cent of the soil surface covered by plant residues in order to increase water infiltration and cut down on soil erosion and runoff. Conservation tillage is an intermediate form of CA since it keeps some soil cover as residue from the previous crop. But some tillage is usually done. It developed as a management system after the "Dust Bowl" of the 1930's in the Mid-West areas of the USA. It was found to reduce erosion by protecting the soil surface from wind and rain.

Direct Planting, Direct Drilling, Plantio Direto and Siembra Directa

These are terms used for ZT in other countries like Australia and South America. They use special equipment (NT drill) to plant seeds directly into crop residues left on the soil surface without preparing a seedbed beforehand.

Direct Seeding

This term is usually associated with growing a rice crop like any other cereal crop without producing seedlings that are then transplanted into the main field. However, it can also be called NT or ZT if the seed are drilled without tillage.

Organic Farming

Organic agriculture does not permit the use of synthetic chemicals to produce plant and animal products, relying instead on the management of soil organic matter and biological processes. In some parts of the world, farms must be inspected and certified before their food products can be sold as organic, indicating that no synthetic chemicals were used in producing them. But organic farming uses the principles of CA to some extent and one objective similar to CA is to maintain and improve **soil health.** Unlike organic farming, CA does allow farmers to apply synthetic chemical fertilisers, fungicides, pesticides and **herbicides.** Many farmers rely on using these to control weed and pest problems, particularly during the early transition years. As soil physical, chemical and biological health improves over time; the use of agrichemicals can be significantly reduced or, in some cases, phased out entirely.

2.6 Strategies for Sustainability

As a result of advancement of civilization and evacuation of the man from the jungle, he entered the farming age. He gradually adopted farming methods using the traditional knowledge gained through generations and observing his environment. The man who made his livelihood by collecting food from the jungle commenced a systematic cultivation of crops and rearing of animals. As the time passed, the man had to face problems with regards to cultivating fruits, vegetables, grains, etc. He learnt by making mistakes and using traditional knowledge on how to prepare the land adequately, perceiving fertility of the soil, controlling pests and diseases of different crops and managing crops providing suitable spaces. The knowledge gained through experience was transferred from generation to generation. The information generated by man using the traditional knowledge and the new knowledge gained through experience had disseminated during several thousand years. When man realised that the land he cultivated had become unfertile after several years of repeated cultivation of monocrop, he shifted his farm to a new land. Such shifting was possible as the lands were available in abundance for cultivation.

2.6.1 High External Input Agriculture (HEIA)

The basic aspect of conventional agriculture was to maintain subsistence level production by using locally available resources. All resources had been naturally recycled and reused without wasting. But due to pressure of increasing population in developing countries steps were taken to expedite food production deviating from the traditional pattern.

The pressure of world population explosion exerted more on the people of developing countries. In order to confront the pressure of world population explosion the farmers as well as the researchers and extensionists were compelled to join the "Seed-Manure" revolution bom with the label **"green gevolution"**. The aim of this "revolution" was to provide food for the increasing population by enhancing the harvest per unit and the intensification of the number of cultivation seasons. Green revolution introduced during the early part of the 6th decade of the 20th century accelerated food production.

Cropping intensity is defined on the basis of number of cultivations per year. If a land is cultivated 2 times a year the cropping intensity is 200 per cent. If it is cultivated only in one season the cropping intensity is 100 per cent. In response to the green revolution practices, introduced in late 1960s our agricultural production increased significantly.

High yielding hybrid seeds which were introduced by green revolution were new to our environment. Farmers had to practice new techniques to get higher production. Due to the fact that hybrid seeds were more sensitive to nutrients, the growers were encouraged to use chemical fertiliser in large quantities as external inputs. As the new crops were foreign to the environment they were susceptible to pests and diseases. Consequently, the necessity arose to apply chemicals, which became an additional burden to growers. Application of chemical fertiliser and pesticides increased the cost of production.

Hybrid varieties were dwarf in nature and could not compete with weeds. Application of herbicides or manual weeding was essential to mitigate the competition between hybrid varieties and weeds. From land preparation to harvesting all agricultural practices, related to hybrid varieties were more labour intensive. Mechanisation was an integral component of green revolution. To increase the working efficiency of the production system machinery like tractors have been introduced. These machines required mineral fuel. The water consumption of new crops was also higher.

Therefore, it was necessary to improve irrigation facilities. External resources were used extensively in this agricultural system. Many of the resources used in high external input agriculture were not recycled. This system failed to add anything to enrich the soil. Many resources were wasted and were beyond the control of the farmers.

As long as external resources are pumped into the system the hybrid varieties-gave maximum yields. Once the provision of resource is stopped the system collapsed and became unproductive. Therefore, to maintain long term sustainability of the system regular application of external inputs was essential.

Advantages of high external input agriculture (HEIA)

1. Agricultural production could be rapidly increased to meet the demand for food for the increasing population.
2. As a result of availability of adequate food stuffs, many problems related to diseases caused by mal-nutrition and deficiency were prevented or reduced.
3. New improved varieties gave yields within a short period of time.
4. Mechanisation solves the problem of labour shortage.
5. Income and profit margins of the products were increased.
6. Productivity of land increased.
7. Increased market facilities for production.

Disadvantages of HEIA

1. Collapse of environmental balance due to lack of bio-diversity by planting a few cash crops.
2. Increase in soil erosion due to constant furrowing by machinery.
3. Dependence on imported machinery, chemical fertiliser, pesticides, hybrid seeds and other inputs.
4. Extensive use of pesticides disturbed the natural mechanism of controlling pest and diseases as the artificial pesticides kill both pests and their natural enemies.
5. Use of artificial agro-chemicals adversely affected the soil pH, cation exchange capacity, soil structure, soil texture and soil organisms. Consequently, the microbial activities of the soil tend to reduce forming dead soil.

6. Although the need for high capital investment, large scale farmers benefited while small scale farmers who were short of capital ran into debt.
7. Neglecting environmental friendly traditional varieties of seeds and their genetic resources faced extinction due to introduction of hybrid varieties. Conventional agricultural knowledge and techniques were neglected and extinguished.
8. Farmers in developing countries had to encounter a series of environmental, soci- economic and political problems as a result of the use of high external input agricultural practices.

Although the production had increased rapidly the demand for excess agricultural production in the market had not commensurately increased. Farmers encountered with the problems in selling of their products and at the same time prices of food items were reduced.

Possible problems of the farmers which would come out in future were not anticipated in planning. Following are some of the short comings of the green revolution:

1. Failure to foresee the fall of prices of surplus agricultural commodities due to increase in production caused by the use of chemical fertiliser, fuel, pesticides, hybrid seeds, etc. which were costly.
2. Failure to realise long-term dependency of the farmers on pesticides and chemical fertiliser and the consequent effect on ecology and human life.
3. Quick response for growth and production and relatively lesser quantities required for one application promoted farmers to apply artificial manure. But the farmers were not aware of long-terms consequences of chemical fertiliser.
4. Efficiency of chemical fertiliser was below the expected level. It is found that 40 to 50 per cent of nitrogen in the chemical fertiliser applied in the tropical countries is wasted. This wastage further aggravates due to factors, such as high rainfall, severe droughts, soil erosion and low content of organic matter in the soil.
5. Chemical fertiliser directly affects on the nutrient balance and the biological activities of the soil. It disturbs the soil structure and expedites decomposition of soil organic matter. Application of acidic nitrogen fertiliser like ammonium sulphate decreases the soil pH and reduces the ability to use phosphorus by plants.

6. Continuous application of nitrogen, phosphorus and potassium fertiliser creates micro-nutrients deficiencies in the soil. Plant display deficiencies of zinc, iron, copper, magnesium, manganese, molybdenum and boron. A reason for this is the barrier imposed by chemical fertiliser for micro-nutrients to enter into the soil. This situation results in the reduction of crop production and susceptibility to pests and diseases.

At present, many advantages gained from high external input agriculture are progressively diminishing and the yield per unit is decreasing. Therefore, the farmers, researchers and extensionists are compelled to contemplate towards a sustainable agricultural system free of high external inputs.

2.6.2 Low External Input Agriculture (LEIA)

Sustainable agriculture is an important element of the overall effort to make human activities compatible with the demands of the earth's ecosystem. While sustainable agriculture is based on long-term goals, but not a specific set of farming approaches, it is usually accompanied by a reduction of purchased inputs in favour of managing on-farm resources. A good example is reliance on biological nitrogen fixation by legumes vs manufactured mineral fertilisers. Low input agriculture (LEIA) is one of several alternative farming systems whose methods are adapted to sustainable agriculture.

Low external input agriculture (LEIA) is based on reduction-but not necessarily elimination-of chemical fertilisers, insecticides and herbicides. Farmers are adapting these practices primarily to reduce costs, but also because they want to minimise impact on the environment or because they perceive future pesticide regulations.

The term **LEIA has been defined** as a production activity that uses synthetic fertilisers or other agro-chemicals below rates commonly recommended. It does not mean elimination of these materials. Yields are maintained through greater emphasis on cultural practices, IPM and utilisation of on-farm resources.

The LEIA means maximum utilisation of local resource with less dependency on external input for agricultural production. LEIA is one of integration form between agriculture with nature and natural process. LEIA can be lead to low external input and sustainable agriculture **(LEISA).** The system of LEIA is suitable for tropical country like

Indonesia, where the natural richness still high and the utilisation of technology is still low. Characteristics of LEIA *vs* LEISA are given in Table 2.4.

Table 2.4 : Characteristics of LEIA and LEISA

LEIA	LEISA
• Farm management which optimises the use of locally available natural and human resources and indigenous technical knowledge to enhance diversity cyclic flow patterns and to build up living soil.	• Farm practices characterised by the use of inorganic fertilisers, low degree of re-cycling and low degree of optimysing nutrient availability.
• Characterised by a conscious drive towards sustainability.	• Lack of conscious drive towards sustainability.
• One approach to sustainable agriculture.	• Refers to main stream farming practices carried out by majority of farmers in the research sites.
• Low cast input approach, high reliance on recycling of on-farm resources.	• Spread and adaptation facilitated by Government agencies.
• Inclusion of biological nitrogen fixing legumes.	
• Emphasis of green manure crops.	

2.6.3 Low External Input and sustainable Agriculture (LEISA)

World food production tremendously increased as a result of **high external input agricultural practices** introduced by the **green revolution.** As a result of the need for excessive capital, unsustainability of the system and negative impact on environment, the farmers had to face many problems. Therefore, development of an agricultural system using lesser external inputs, less expensive and environment friendly has become a need of many countries. Agricultural practices with lesser inputs have been developed by integrating selected traditional basic principles with new technological knowledge.

This system which is abbreviated as **LEISA** is frequently used for low external input and sustainable agriculture. The term **LEISA is an integrated agricultural system** consisting of bio-dynamics, environmental, natural, organic and regeneration sub-system. Low external input sustainable agriculture (LEISA) relies mostly on the inputs from the local farm, village or region and deliberate action is taken to ensure sustainability. LEISA is based on the following ecological principles:

- Creating a favourable condition for growth and sustenance of plant by stimulating soil micro-organisms as far as possible and adding organic matter sufficiently.
- Maintaining nutrient content at optimum level assuring the balance of nutrients in the soil by nitrogen fixation, utilisation of nutrients available in the deep soil layers, promotion of recycling process and addition of external fertiliser as and when necessary to complement deficient nutrients.
- Controlling the micro-climatic conditions to minimise loss of resources, due to sunlight, air and water. Use of biological and mechanical methods to prevent soil erosion.
- Minimising loss of resources caused by pests and diseases. Integration of pest control methods giving priority to natural biological control of pests by natural enemies on the principle that prevention is better than eradication.
- Promoting bio-diversity and complexity. Stimulating synergetic and symbiotic conditions between plants/plants and plants/animals.

In **the high external input agriculture (HEIA),** production for the far-off markets necessitated use of external inputs like chemical fertilisers, hybrid seeds, pesticides, irrigation, etc. Increased dependency on high cost external inputs in agriculture also made farmers to depend on external credit on a regular basis. Cultivation of cash crops like cotton and tobacco, also led to scarcity of fodder. This resulted in farmers giving up animal husbandry, thereby resulting in acute scarcity of **farmyard manure** and making the use of chemical fertilisers inevitable.

Adoptions of modem technologies in agriculture like tractors and pump sets have resulted in the neglect of draught animals. Even the livestock production has been totally changed into industrial type of production from backyard system. Animals provide manure, food and income and are used for cultivation and transport economically. Different animals can be fed on farm wastes efficiently, which provide financial security at the time of distress. In Indian culture, cattle are treated as an integral part of the family. After undergoing the huge losses from adapting crossbred cows like Holstein, Frisien and Jersey, farmers are fast changing towards indigenous cattle, since their maintenance is cheaper and the male calves can be used for cultivation and transport.

Agriculture in India was an integrated cultivation of crops, animals and trees to meet most of the family and community needs rather than market. Trees played an important role in providing **green manure,** fodder, fruits, fuel and timber besides conserving soil, water and hosting beneficial insects and birds. Now, it is necessary to revive the traditional knowledge on seed selection and preservation to bring back the self-reliance and seed availability at the time of sowing.

Similarly, rural population is fast changing their medical treatments since synthetic drugs have become more expensive and found to create side effects. They now started growing medicinal plants like neem, ashwagandha, asparagus, aloevera, adathoda, etc. for medication of both humans and their livestock.

The Government's concern of impending scarcity of fossil fuel resources is an encouraging factor for promotion of **ecological agriculture.** However, the fear of lower food production and availability of biomass is coming in the way of progressing towards **organic farming.** The goal of **LEISA,** therefore, should be based on economic interests by popularising production of seeds, **vermicompost, botanical pesticides,** etc. A brief account of LEIA vis- a-vis HEIA is presented in Table 2.5.

In agricultural activities, the foregoing principles could be utilised by means of different technologies and methods.

The objective of LEISA system is to maintain the agricultural production at an optimum level using less external inputs in eco-friendly environment. To achieve this objective, the LEISA practices concentrated heavily on the following factors:

1. Maintaining a living soil.
2. Creating bio-diversity.
3. Re-cycling of resources.
4. Natural pest management.

Maintaining a Living Soil

- Growing legumes to fix nitrogen from the air and provide it to the following crop.
- Feeding the soil with as much organic matter as possible through green manure, compost, cover crops, returning of non-toxic organic wastes and agro-forestry.
- Keeping the soil covered at all times with mulch or cover crops.
- No or reduced tillage which enhances water availability and soil conservation.

Table 2.5 : LEIA *vis-a-vis* HEIA.

S No	Characteristics of HEIA	Characteristics of LEIA
1.	Farming pattern depends heavily on external and chemical inputs.	LEIA relies on the optimal use of natural processes. Although yields have increased substantially, contributing to raising total production, farmers and the environment have had to pay the price for keeping up with this development.
2.	Focus of agricultural development and research has mainly been on maximising yields coupled with increasing specialisation of production.	Focus is on the sustainability of farming system.
3.	There is great damage to the environment.	Environmentally sound and that have the potential to contribute to the long-term sustainability of agriculture.
4.	Continuing drop in prices of farm produce and the rising costs of agricultural inputs have made farming increasingly unprofitable.	Greater emphasis is on the long-term sustenance and balance between the profit and livelihood.
5.	HEIA depends on the higher production and profit, without consideration of the local needs and local market.	Sustainable ecological practices depend largely on local agro-ecological conditions and on local socio-economic circumstances, as well as on farmers' individual needs and aspirations.
6.	Primarily one or two commodity driven development, lack of diversity in the farming practices; as a result, there is greater risk of failure and price fluctuation. The number of products and commodities are very minimum.	One way of LEIA is to diversification of farms; with a range of crops and/or animals, farmers will suffer less from price fluctuations or drops in yield of single crops. Maintaining diversity will also provide a farm family with a range of products to eat or sell throughout a large part of the year.
7.	Under HEIA system, soil quality deteriorates and there is resurgence of pests, lack of resilience in the soil-plant system.	LEIA maintains a healthy soil, recycling nutrients on the farm, and utilising approaches, such as integrated pest management (IPM).
8.	In HEIA, there is lack of use of indigenous technologies.	Best bet technologies, for example, soil and water conservation (terraces, ditches, and vegetation strips on sloping land), better timing of operations, improved crop spacing and densities, manure or compost and water application based on local conditions.

- Cultivating a range of crops and animals to reduce risks of disease and pest outbreaks, maintain a balanced nutrient supply and provide resilience.
- Planting trees on contours or making terraces to prevent soil erosion by wind or water.

Creation of Bio-diversity

- Intercropping in time and space: planting different crops together in different combinations or formations or in sequence can optimise the use of available resources and reduce the pressure of pests.
- Different plant species can also be used to support the ecological functioning of the whole farm system: examples are trees or bushes for windbreaks, flowering plants which provide food and habitat for beneficial insects that help control pests, shade trees for light-sensitive plants, trees to provide green manure and fuelwood.
- Integration of different crops or weeds with animals to better utilises resources, for example fish in rice fields, integrated crop-chicken-fish system and other combinations of crops and animals.

Recycling of Resources

Natural forest ecosystem does not use external resources other than sunlight, air and water. In such system, the wastage of resources is minimised and instead the resources are being used again and again. Leaves, fruits, flowers, trunks and other vegetative and reproductive parts of plants together, with animal wastes and dead bodies are decomposed to form humus. Humus provides nutrients to plants. If there are no external influences, this process will continue forever. In a paddy field, the paddy straw can be re-cycled to give nutrients. Promotion of recycling process will help to maintain the sustainability of the farming system.

Natural Crop Protection Methods

A series of actions such as reduction, prevention, extermination and controlling of pgst attack can be used. Seeds and planting materials should be treated to destroy pathogens before planting them in the field. Sometimes, hot water treatment will be sufficient for this purpose. Use of wood-ash and sun-drying also possible. It is advisable to use farm implements hygienically as there are possibilities of transmitting pests and diseases through them. Infested fruits, plants leave **ratoon crops** and host plants of pests and diseases should be kept away from the farm. Some promising LEISA techniques and practices are indicated below:

1. Nutrient management.
2. Integrated pest management (IPM).
3. Crop residue management and conservation tillage.
4. Converting farm wastes into useful organic manure.
5. Green manuring.
6. Cover crops.
7. Alley cropping.
8. Multiple cropping.
9. Intercropping.
10. Relay cropping.
11. Practices for land degradation.
12. Organic manuring.
13. Mulching.
14. Windbreaks.
15. Water harvesting.
16. Stripcropping.
17. Companion cropping.
18. Decoy crops.
19. Contour farming.
20. Integrated crop-livestock-fish farming, etc.

□□□

7

Integrated Farming System

All over the world, during the last four decades, grain yields went up at spectacular rates during the **green revolution** and individual levels of production in animals followed similar trend. Ensuring food security for the fast growing global population is a mammoth challenge for the present agricultural production system. Shrinking average farm size in India and financial constraints for higher investment in agriculture due to 80 per cent farm families belonging to small and marginal farmer categories further heighten the challenge. For securing food and nutrition security for sizable population, productivity enhancement may provide a vital solution. This involves adoption of scientific agronomic practices and technologies which promise an augmentation of the productive capacity of traditional agricultural system. Agronomic practices such as the liberal use of inorganic pesticides and fertilisers during the twentieth century enhanced productivity significantly but undesirable **environmental degradation** accompanied by increased operational costs in agriculture raised concerns about economic feasibility and sustainability. Unsustainable farming leads to **environmental pollution** and threatens the livelihood of millions of small farm holders. Strengthening agricultural production systems for greater **sustainability** and higher economic returns is a vital process for increasing income and food and nutrition security in developing countries.

Therefore, **integrated farming system (IFS)** is a multi-disciplinary whole farm approach and appears effective in solving the problems of small and marginal farmers. The approach aims at increasing income and employment from small-holding by integrating various farm enterprises and **re-cycling crop residues** and by products within the farm itself. Farmers need to be assured of regular income for living at least above poverty line. Progress in production or steady growth in output is necessary to face the challenges posed by present economic, political and technological environment. In this context, **farming system approach** is one of the important solutions to face this peculiar situation as in this approach different enterprises can be carefully undertaken and the location specific system are developed based on available resources which will result in **sustainable development.**

Emergence of integrated farming system has enabled us to develop a framework for an alternative development model to improve the feasibility of small sized farming operations in relation to larger ones. **Integrated farming system** (or integrated agriculture) is a commonly and broadly used word to explain a more integrated approach to farming as compared to **monoculture** approaches. It refers to agricultural system that integrates livestock and crop production or integrates fish and livestock and may sometimes be known as **integrated bio-system.** In this system, an inter-related set of enterprises used so that the "waste" from one component becomes an input for another part of the system, which reduces cost and improves production and/or income. *The IFS works as a system of systems.* It ensures that wastes from one form of agriculture become a resource for another form. Since it utilises wastes as resources, we not only eliminate wastes but we also ensure overall increase in productivity for the whole agricultural system. As such, IFS is the only way through which the target could be achieved.

Prof George Chan, working with a network of **polyculture** farming pioneers, began refining integrated farming practices that had already been developed in south-east Asia in the 1960 through the 1980s, building on traditional night soil farming practice ("A Primer on Integrated Farming System - One Village Foundation, USA). In China, programmes embracing this form of integrated farming have been successful in demonstrating how intensive growing system can use organic and sustainable farming practices, while providing high agriculture yields.

"Integrated Food and Waste Management System" (IF&WMS) which was developed by Prof. Chan is one version of an IFS. He introduced this concept at the Montfort Boy Farm in Fiji, a vocational school that now serves as a model for the students to replicate in their villages (A Primer on Integrated Farming System). Today, there are numerous IF&WMS or IFS models. These systems combine livestock, aquaculture, agriculture and **agro-industry** in an expanded symbiotic or synergistic system, so that the wastes of one process become the input for other processes, with or without treatment to provide the means of production, such as energy, fertiliser and feed for optimum productivity at minimum costs. The concepts associated with IFS are practiced by numerous farmers throughout the globe. A common characteristic of these systems is that they have a combination of crop and livestock enterprises and in some cases may include combinations of aquaculture and trees. It is a component of farming system which takes into account the concepts of minimising risk, increasing total production and profits by lowering external inputs through re-cycling and improving the utilisation of organic wastes and crop residues. In this respect, integration usually occurs when outputs (usually by-products) of one enterprise are used as inputs by another within the context of the farming system. Difference between **mixed farming** and **integrated farming** is that enterprises in the integrated farming system interact eco-biologically, in space and time, are mutually supportive and depend on each other.

The IF (integrated farming) is system oriented multi-component agriculture. It views the farming in a holistic manner emphasizing the interactions between components. Given the multi-biological components as its central part, integrated farming is a complex biological- social-economic system. Final goal of management is to seek for the integrated effects of the whole system, not the effects of its individual compartment (Fig. 3.1).

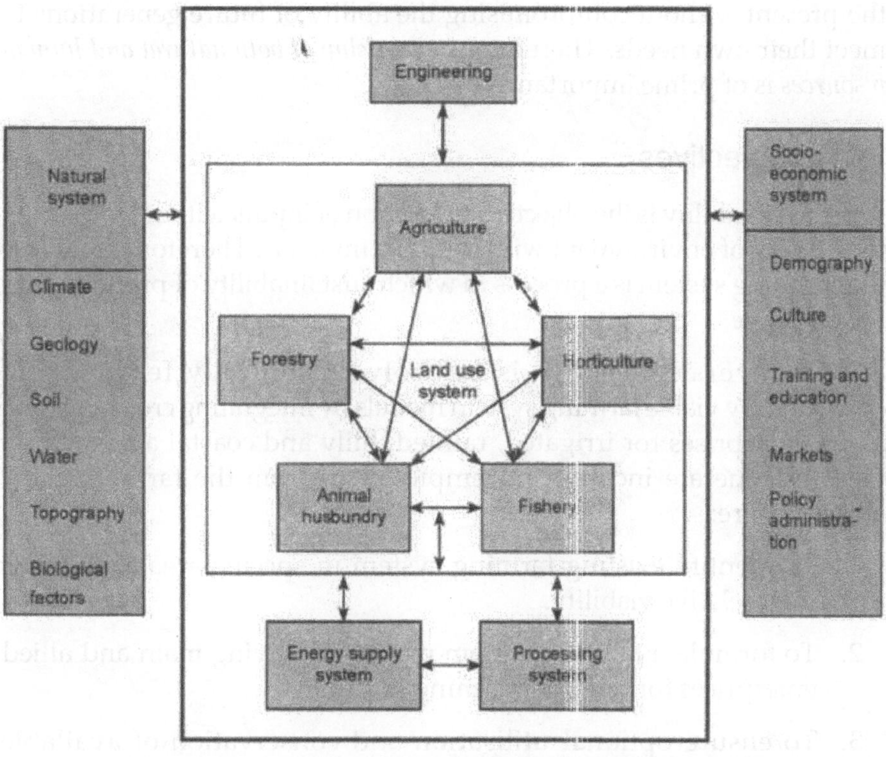

Fig. 3.1: Structure of an integrated farming system (IFS).

Integrated farming system may be defined as a set of agricultural activities organised into a functional unit to profitably harness the solar energy while preserving land productivity, environmental quality and maintaining the desirable level of **biological diversity** and ecological stability. Integration of different agricultural allied enterprises with crop activity as base would provide ways to reuse and recycle produce/ waste material of one component as input in the other linked component and to reduce the cost of production of the economic produce of the component two and finally to enhance the net income of the farm as a whole.

Integrated farming system, can also be, defined as a new type of system, in which multiple agricultural productions and development are guided, organised and managed in the light of ecological-economic principles and by using the system engineering approach.

3.1 Objectives and Characteristics

Sustainability rests on the principle that we must meet the needs of the present without compromising the ability of future generations to meet their own needs. Therefore, *stewardship of both natural and human resources* is of prime importance.

3.1.1. Objectives

Sustainability is the objective utilisation of inputs without impairing the quality of environment with which it interacts. Therefore, it is clear that farming system is a process in which sustainability of production is the objective.

The overall objective is to evolve technically feasible and economically viable farming system models by integrating cropping with allied enterprises for irrigated, rainfed, hilly and coastal areas with a view to generate income and employment from the farm. Specific objectives are:

1. To identify existing farming system in specific area and assess their relative viability.
2. To formulate farming system models involving main and allied enterprises for different farming situations.
3. To ensure optional utilisation and conservation of available resources and effective re-cycling of farm residues within system.

4. To maintain sustainable production system without damaging resources base of environment.
5. To raise overall profitability of farm household by complementing main allied enterprises with each other.
6. Analysis of existing farming system in terms of resource use efficiency, production and productivity, income, employment and sustainability across different agro-climatic zones.
7. Evaluation and identification of farming system through participatory approach that ensures threshold level of income for the livelihood security.
8. Development and evaluation of synergic effects and their actions associated with different farming systems.
9. Developing appropriate institutional and market linkage including value addition for enhancing sustainability.
10. Imparting training and capacity building of various stakeholders on IFS.

3.1.2 Farming Systems and Their Characteristics

Goal orientation of a farm is taken to be an organised decision-making unit in which crop and/or livestock production is carried out with the purpose of satisfying the farmer's goals. On large scale, market production and profits are the main objectives whereas for the small-holder farmer who farm most of the tropics, the farm is a multi-objective system in providing food for the household, raw materials for building huts, accumulation of capital in the form of animals or plantations and accumulation of wealth.

The farm as a system has a boundary that separates the system from the environment. The system embraces all workers and resources (elements of the system) which are under the management of the farmer.

Each individual farm has its own specific characteristics, which arise from variations in resource endowments and family circumstances. The household, its resources and the resource flows and interactions at this individual farm level are together referred to as a farm system. A **farming system** is defined as a population of individual farm system that has broadly similar resource bases, enterprise patterns, household livelihoods and constraints and for which similar development strategies and interventions would be appropriate.

In attempting to combat hunger and poverty, developing countries face the challenges of identifying specific agricultural and rural development needs and opportunities and focusing investment in those areas where the greatest impact on food insecurity and poverty could be achieved. Delineation of farming system provides a useful framework within which appropriate agricultural development strategies and interventions can be determined, as by definition, they group farm households with similar characteristics and constraints. Only a limited number of systems are delineated within each region, leading inevitably to a considerable degree of heterogeneity within any single system. However, the alternative of identifying numerous, discrete, micro-level farming system in each developing region would detract from the overall impact of the analysis.

Classification of the farming system (1.7.4) has been based on a number of key factors, including:

1. The available natural resource base.
2. The dominant pattern of farm activities and household livelihoods, including relationship to markets.
3. The intensity of production activities.

These criteria have been applied to each of the six main regions of the developing world. The exercise resulted in the identification of 72 farming systems with an average agricultural population of about 40 M inhabitants (FAO). Based on these criteria, eight broad categories of farming systems have been distinguished:

1. Irrigated farming system, embracing a broad range of food and cash crop production.
2. Wetland rice-based farming system, dependent upon seasonal rains supplemented by irrigation.
3. Rainfed farming system in humid areas characterised by specific dominant crops or mixed crop-livestock system.
4. Rainfed farming system in steep and highland areas, which are often, mixed crop- livestock system.
5. Rainfed farming system in dry or cold low potential areas, with mixed crop-livestock and pastoral system merging into system with very low current productivity or potential because of extreme aridity or cold.

6. Dualistic (mixed large commercial and small holders) farming system, across a variety of ecologies and with diverse production patterns.
7. Coastal artisanal fishing system, which often incorporate mixed farming elements.
8. Urban-based farming system, typically focused on horticultural and livestock production.

Except for the dualistic system, the systems within each category are dominated by smallholder agriculture. The names chosen for individual farming system reflect the eight categories outlined above. They also reflect key distinguishing attributes, notably:

1. Water resource availability – irrigated, rainfed, moist, dry.
2. Climate – tropical, temperate, cold.
3. Landscape relief/altitude – highland, lowland.
4. Farm size – large scale.
5. Production intensity – intensive, extensive, sparse.
6. Dominant livelihood source – root crop, maize, tree crop, artisanal fishing, pastoral.
7. Dual crop livelihoods – cereal-root, rice-wheat (note that crop-livestock integration is denoted by the term mixed).
8. Location – forest-based, coastal, urban based.

3.2 Components of Integrated Farming System

Following are the major components of farming system across agro-ecological situations in the country:

Agriculture	Kitchen garden
Fish farming	Poultry
Horticulture	Fodder production
Duck rearing	Goat rearing
Forestry	Nursery
Pigeon rearing	Sheep rearing
Mushroom cultivation	Seed production
Sericulture	Piggery
Azolla rearing	Vermiculture
Dairy	Bio-gas

Different IFS models are:

1. Crop-livestock-forestry farming system.
2. Crop-fish-poultry farming system.

3. Crop-livestock-poultry-fishery farming system.
4. Rice + fish only.
5. Fish only.
6. Nilembium + fish, etc.

In general, all the above components are grouped into *three important components of a farming system* (whole farm business):

1. *Crop component* (cereals, pulses, oilseeds, sugar, fiber, vegetables, fruits, agro-forestry, etc.) usually taken up in the main crop fields.
2. *Animal component* (cattle, goat, sheep, etc.) usually taken up within the homestead.
3. *Homestead farming* (bio-gas, post-harvest value added products, grinding, splitting of pulses) which include other allied activities taken up within the homestead.

Occasionally, soil, water and labour are also included under the components of a farming system.

Each of the above components may have one or more of several activities or processes. Thus selection, production/cultivation and /or rearing of the activities within a farm will determine the nature of farming system being practiced in any situation. Nevertheless, the kind of activities (crop or animals) included under each component depend on their suitability, adaptability, marketability and ability to satisfy the needs of a farm family. Experience of the farmer and his predecessors (father and forefather) over years enable them to select the activities under each system. In fact, the farmer has been experimenting with the crops and the crop varieties or species and animal types or breeds knowingly or unknowingly over years before finally incorporating these in the farm plan. Therefore, basically it is a case of simulation and experimenting with crops and animals before incorporating them in any system. Once a system is identified, its adoption will primarily depend on availability of different types of limited resources at the disposal of the farmer.

Advantages of integrated farming systems

1. **Re-cycling and utilisation of other available resources in the farm:** There is effective re-cycling of waste material in farming system. Farming is a stochastic dynamic biological and open system with human and social involvement. It specifically refers to crop-combination or enterprise-mix in which the products

and/ or the by-products of one enterprise serve as the input for the production of other enterprise. Farming system includes agricultural crops, fruit trees, forest trees, forages, grasses and animals, which have potential to contribute to improve resource use and management.

2. **Maximum possible return and profitability:** Use of waste material of one component in the other at the least cost reduces cost of production and net profit is increased. Due to interaction of enterprises like crops, eggs, milk, mushroom, honey, cocoons silkworm farming, there is flow of money to the farmer round the year.
3. **Create adequate employment opportunities:** Combining crop with livestock enterprises would increase the labour requirement and help in reducing the problems of under employment. IFS provide enough scope to employ family labour round the year.
4. **Productivity:** IFS provides an opportunity to increase economic yield per unit area per unit time by virtue of intensification of crop and allied enterprises.
5. **Potentiality or sustainability:** Organic supplementation through effective utilisation of by-products of linked component provides an opportunity to sustain the potentiality of production base for much longer periods.
6. **Balanced food:** The linked components of varied nature enable to produce different sources of nutrition.
7. **Environmental safety:** In IFS, waste materials are effectively re-cycled by linking appropriate components, thus minimise environment pollution.
8. **Adoption of new technology:** Resourceful farmers (big farmer) fully utilise the technology for profits. Money flow round the year gives an inducement to the small/marginal farmers to go for the adoption of latest technologies.
9. **Saving energy:** Organic wastes available in the system can be utilised to generate bio-gas. This alternative source, thus, reduces dependence on fossil energy sources j within short time.
10. **Meeting fodder crisis:** Plantation of perennial legume fodder trees on field borders fixes the atmospheric nitrogen. These practices will greatly relieve the problem of non- availability of quality fodder to the animal component linked.

11. **Solving fuel and timber crisis:** By linking agro-forestry appropriately, production level of fuel and industrial wood can be enhanced without any effect on crop. This will also greatly reduce deforestation, thus preserving our natural ecosystem.
12. **Agro-industries:** When one of produces in farming system is increased to commercial level, the surplus lead to the development of allied agro-industries.

3.3 Site Specific Development of IFS Model for Different Agro-Climatic Zones

Farming enterprises include crops/cropping system, poultry, fish, piggery, sericulture, I mushroom, agro-forestry, bio-gas, vermicompost etc. A combination of one or more enterprises with cropping when carefully chosen planned and executed, give greater dividends than I single enterprise(s) specially for small and marginal farmers. Farm as a unit to be considered and planned for effective integration of the enterprises to be combined with crop production activities. In the IFS, it is always emphasised to combine cropping with other enterprises/ activities.

3.3.1 Determinants of Farming System

The key categories of determinants influencing farming system are briefly presented.

Natural Resources and Climate

Interaction of natural resources, climate and population determines the physical basis for farming system. Increased variability of climate and thus agricultural productivity substantially increases the risk faced by farmers with the concomitant reduction in investment and input use. Certain soil and water regimes are suitable only for given type of crops. Similarly, some of the physical and geographical features e.g drainage characteristics, elevations and slopes as well as climatic factors e.g. total rainfall and its distribution, minimum and maximum temperature, humidity and intensity of sunlight, etc. are other factors which have to be taken into considerations while making decision with respect to selection of enterprise for a farming system.

Science and Technology

Investment in agricultural science and technology has expanded rapidly during the last four decades. During this period, major technical and institutional reforms occurred, which shaped the pattern of technology development and dissemination. Research driven growth in developing countries has been **green revolution,** where it achieved considerable achievement in the field of foodgrain production and for this the policy and other aspects supported the farming system for such achievement. Research has been focussed principally upon intensifying crop and livestock production. There has been for less research on integrated technologies for diversifying the livelihoods of small farmers in developing countries and increasing the sustainability of land use. Despite these weaknesses, the natural and global research agenda is gradually moving from a focus on individual crop performance to a growing acceptance of the importance of increased system productivity. There has been emphasis in recent agriculture of targeting technologies towards women farmers and poorer households.

Trade Liberalisation and Market Development

Markets have a critical role to play in agricultural development as they form the linkages between farm, rural and urban economics upon which the development processes depend. As a result of reduction of impediments to international trade and investment, the process of trade liberalisation is already generating changes in the structure of production at all levels- including small holder-farming system in many developing countries. Not only the market development is accelerating, but patterns of production and natural resources usage are also changing profoundly in response to market forces.

Availability of new production, post-harvest and transport technologies will also change demand patterns due to delivery of new products or established products in new forms to markets, where they have been previously unattainable.

Policies, Institutions and Public Goods

Development of dynamic farming system requires conducive policy environment. Moreover, establishment of farm-rural-urban linkages requires effective demand. Policy-makers have increasingly shifted their attention to the potential to increase the efficiency of service delivery through the restructuring of institutions. Production incentives have dramatic effect on farming system. Policies on land ownership, water management and taxation reform, etc. have a great bearing on types of farming system in a region or area.

Information and Human Capital

Evolution of farming system based upon increasing specialisation (large scale broiler units) or integrated intensification (rice-fish-ducks) requires extra knowledge on the part of farm operators. Need for better information and enhanced human capital has also increased, as production systems have become more integrated with regional, national and international market system.

Lack of education, information and training is frequently a key limiting factor to smallholder development. Many observers anticipated an information revolution i.e. bridge gap of knowledge between scientists and farmers will be very key factor for agricultural growth of these small farmers.

Similarly, better access to credit, land and extension services would enable women to make an even greater contribution to eliminating rural hunger and poverty. As gender bias is progressively eliminated during coming years - often in the face of severe cultural and religious barriers, productivity within many farming system will be transformed.

Indigenous Technical Knowledge (ITK)

Indigenous technical knowledge is the knowledge that people in a given community has developed over times and continues to develop. It is based on experience, often tested over long period of use, adapted to local culture and environment, dynamic and changing and lays emphasis on minimising risks rather than maximising profits. The ITK covers a wide spectrum – soil-water and nutrient management; pasture and fodder management; crop cultivation; plant protection; farm equipment, farm power, post-harvest preservation and management; agro-forestry; bio-diversity conservation and also exploitation; animal rearing and health care; animal products preservation and management; fisheries and fish preservation and ethnic foods and homestead management. Thus, the ITK of a farmer has a great influence in managing the farm and farming system.

3.3.2 Criteria For Enterprise Selection

The basic points that are to be considered while choosing appropriate enterprise in IFS are:

1. Soil and climatic features of an area/locality.
2. Social status of the family and social customs prevailing in the locality.

3. Economical condition of the farmer (return/income from the existing farming system).
4. Resource availability at farm and present level of utilisation of resources.
5. Economics of proposed IFS and credit facilities.
6. Farmer's managerial skills.
7. Household demand.
8. Institutional infrastructure and technological know-how.
9. Market facilities.

Priority should be given to improve the present status of different components of the existing farming system, should have better compatibility with prevalent farming system, nil to very less dependence of input from outside, high risk bearing and capable to generate more per day income and employment.

3.3.3 Interactions and Linkages

The IFS deals with utilisation of wastes and residues. It may be possible to reach the same level of yield with proportionately less input in the integrated farming and yield would be interestingly more sustainable because the waste of one enterprise becomes the input of another leaving almost nothing to pollute the environment or to degrade the resource base. To put this concept to practice efficiently, it is necessary to study linkages and complementarities of different enterprises in various farming system. Knowledge of linkages and complementarities will help to develop farming system (integrated farming) in which the waste of one enterprise is more efficiently used as inputs in another with the system (Fig 3.2).

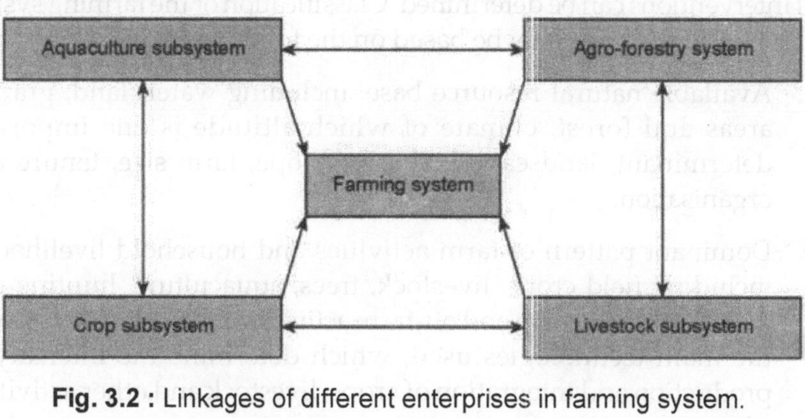

Fig. 3.2 : Linkages of different enterprises in farming system.

3.3.4 Integration of Enterprises

In agriculture, crop husbandry is the main activity. Income obtained from cropping is hardly sufficient to sustain the farm family throughout the year. Assured regular cashflow is possible, when crop is combined with other enterprises. Judicious combination of enterprises, keeping in view of the environmental condition of locality will pay greater dividends. At the same time, it will also promote effective re-cycling of residues/wastes.

A farm manager is often confronted with the problems as to what enterprises to select and the level at which each enterprises should be taken up. How far he can go far or should go in combine enterprises with another depends partly on the inter-relationships, between different enterprises and the prizes of products and inputs.

Types of enterprises include:

1. Independent enterprises.
2. Competitive enterprises.
3. Supplementary enterprises.
4. Complementary enterprises.

Further details on the above types are beyond the scope of this publication.

3.3.5 Integrated Farming System for Different Agro-Ecosystem

Delineation of the major farming system provides a useful framework within which appropriate agricultural development strategies and interventions can be determined. Classification of the farming system of developing regions may be based on the following criteria:

1. Available natural resource base, including water, land, grazing areas and forest, climate of which altitude is one important determinant, landscape including slope, farm size, tenure and organisation.
2. Dominant pattern of farm activities and household livelihoods, including field crops, livestock, trees, aquaculture, hunting and gathering, processing and off-farm activities and taking into account the main technologies used, which determine the intensity of production and integration of crops, livestock and other activities.

Based on these criteria, eight broad categories of farming system have been distinguished. Readers are advised to refer 3.1.2 for further details in this regard.

IFS for different agro-climatic zones (India)

1. ***High altitude cold deserts:*** Pastures with forestry, goats, angora rabbits and limited settled agricultural crops like millets, wheat, barley and fodder.
2. ***Arid and desert region:*** Centering mainly in animal husbandary with the camels, sheep and goat and with moderate cropping components involving pearlmillet, wheat, pulses, gram and fodder crops.
3. ***Western and central Himalayas:*** Horticultural crops as major component have less intensive agriculture mainly on the hill terraces and slopes with maize, rice, wheat, pulses and fodder crops.
4. ***Eastern Himalayas:*** Primitive crop husbandary with rice, millets, pulses, etc. Agro-forestry system is also common. Piggery and poultry are the chief livestock activity.
5. ***Indo-Gangetic plains:*** Intensive crop husbandary involving rice, wheat, maize, mustard, pulses and livestock inclusive of dairy cattle and buffaloes.
6. ***Central and southern highlands:*** Cotton, sorghum, millet, pulses with dairy cattle, sheep and goats and poultry are the secondary livestock and animal husbandary enterprises.
7. ***Western Ghats:*** Major activity on plantation crops, cultivation of rice and pulses are the secondary agricultural activity. Cattle, sheep and goats are the livestock components which in most parts are maintained as large herds and allowed to range.
8. ***Delta and Coastal plains:*** Rice cultivation along with fish culture, poultry and piggery enterprises, capture fisheries of the marine ecosystem is a specialised enterprise and does not mix with cropping activity.

3.4 Resource Use Efficiency and Optimisation Techniques

The objective of any farming research and development effort is to improve the efficiency and productivity of the use of basic resources in the production process. In order to determine the expected benefits, losses and other implications of a proposed change, it is necessary to evaluate the management and performance of both the existing production systems and the recommended improvements. Consideration of economic factors together with bio-physical factors provides a logical framework for comparing traditional and alternative systems. Economics provides a rational basis for making decisions in allocating scare resources among various options to achieve competing goals.

3.4.1. Evaluation of FSR/IFS

- **Productivity per unit area:** To estimate the productivity of a component and compare with the crop component expressed in terms of **equivalent crop yield.** Further, the production estimation itself varies among the inter-linked animal component in IFS (rice based farming system). Productivity, in term of grain yield, can be recorded and expressed as kg of **grain equivalent yield (GEY):**

$$GEY = \frac{\text{Productivity of component intercrop (Kg)} \times \text{Cost of } \frac{\text{Component}}{\text{intercrop}} (\text{Rs. kg}^{-1})}{\text{Cost of main crop (Rs. kg}^{-1})}$$

- **Economic analysis:** Parameters like cost of cultivation/ production, gross and net returns and per day return can be worked out and expressed as Rs. ha^{-1}.

- **Employment generation:** Labour required for various activities in crop production given as man-days ha^{-1} $year^{-1}$ (A man working for 8 hours in a day is considered as one man-day; A woman working for the same period is treated as 2/3 man day and computed to man-days).

- **Productivity of livestock components:** Milk (per day or lactation), dung, urine, etc. *Fisheries:* Fish weight recorded at harvest and expressed as kg unit $area^{-1}$.

  Poultry: Egg production per day from birds and expressed as total number per month per year from the unit.

  Pigeon: Growth rate at monthly interval and weight at the time of disposal recorded and expressed as kg $unit^{-1}$.

- **Mushroom:** Yield per day and total yield per year from the unit.

- **Water requirement:** Water requirement for varying component linkages in the IFS expressed in ha-mm.

- **Residue addition:** The quantity of residue available from each component (kg). Potential residue addition in terms of N, P and K.
- **Energy efficiency:** Energy input and outputs were worked out for individual components based on the input and output energies and energy efficiency.
- **Nutritive value:** Nutritive value in terms of carbohydrates, proteins and fat (kg).

3.4.2. Resource Use Efficiency

Efficiency in any system is an expression of obtainable output with the addition of unit amount of input. The ratio of energy intake and energy of the produced biomass *i.e,* of input and output is called **ecological efficiency.** This can be studied at any tropic level.

Resource use efficiency in agriculture is defined to include the concepts of technical efficiency, allocative efficiency and economic efficiency. Public investment, subsidies and credit for agriculture are used in an efficient manner. There are large scale inter-regional as well as inter-farm variations in **factor productivity** due to varying influence of different factors in different regions. A number of management factors, such as timeliness and method of sowing, transplanting, irrigation and application of right doses of inputs and input mix play an important role in influencing inter-farm variation in crop productivity. Growing marginalisation and fragmentation of land holdings coupled with rising incidence of informal tenancies and poor rural infrastructure, such as road, electricity, markets and education affect factor productivity. Availability of good quality irrigation water coupled with flexibility of irrigation and drainage system and appropriate methods of application as well as pricing of irrigation water is crucial for sustainable use of land and water resources.

Three types of efficiencies have been identified in the literature: technical efficiency, allocative efficiency and overall or economic efficiency as indicated above. **Technical efficiency** is the ability of a firm to produce a given level of output with minimum quantity of inputs under a given technology. **Allocative efficiency** is a measure of the degree of success in achieving the best combination of different inputs in producing a specific level of output considering the relative prices of these inputs. **Economic efficiency** is a product of technical and allocative efficiency. In one sense, the efficiency of a firm is its success in producing as large an amount of output as possible from given sets of inputs. Maximum efficiency of a firm is attained, when it becomes impossible to reshuffle a given resource combination without decreasing the total output

Detailed procedure for calculating resource use efficiency is beyond the scope of this publication. However, resource use efficiencies for commonly used farm inputs in crop production are briefly presented.

Cultivated Land Utilisation Index (CLUI)

Cultivated land utilisation index is calculated by summing the products of land area to each crop, multiplied by the actual duration of that crop divided by the total cultivated land times 365 days.

$$\text{CLUI} = \frac{\sum_{i=1}^{n} a_i d_i}{A \times 365}$$

where, n = Total number of crops.

a_i, = Area occupied by the *i*th crop.

d_i = Days that the *i*th crop occupied.

A = Total cultivated land area available for 365 days.

The CLUI can be expressed as a fraction or percentage. This gives an idea about how the land area has been put into use. If the index is 1 (100%), it shows that the land has been left fallow and more than 1, tells the specification of **inter-cropping** and **relay cropping.** Limitation of CLUI is its inability to consider the land temporarily available to the farmer for cultivation.

Fertiliser Use Efficiency (FUE)

The FUE is the output of any crop per unit of the nutrient applied under a specified set of soil and climatic conditions. The NUE/FUE can be expressed in several ways:

1. *Partial factor productivity* (PFP, kg crop yield per kg input applied).
2. *Agronomic efficiency* (AE, kg crop yield increase per kg nutrient applied).
3. *Apparent recovery efficiency* (ARE, kg nutrient taken up per kg nutrient applied).
4. *Physiological efficiency* (PE, kg yield increase per kg nutrient taken up).

5. *Crop removal efficiency* (removal of nutrient in harvested crop as per cent of nutrient applied) is also commonly used to explain nutrient efficiency.

Energy Efficiency

Energy efficiency

$$= \text{Energy output (MJ ha}^{-1}\text{)/Energy input (MJ ha}^{-1}\text{)}.$$

Net energy (MJ ha^{-1})

$$= \text{Energy output (MJ ha}^{-1}\text{) - Energy input (MJ ha}^{-1}\text{)}.$$

$$\text{Energy productivity} = \frac{\text{kg}}{\text{Mj}} \text{- Output} \left(\frac{\text{Grain+by product, kg ha}^{-1}}{\text{Energy output (Mj ha}^{-1}\text{)}} \right)$$

$$\text{Energy intensity (in physical terms MJ ha}^{-1}\text{)} = \frac{\text{Energy output (Mj ha}^{-1}\text{)}}{\text{Grain+by product, kg ha}^{-1}}$$

Energy intensity (in economic terms MJ Rs.$^{-1}$)

$$= \frac{\text{Energy output (Mj ha}^{-1}\text{)}}{\text{Cost of cultivation (Rs. ha}^{-1}\text{)}}$$

Water Use Efficiency

Crop water use efficiency: Crop water use efficiency is a ratio between marketable crop yield and water used by the crop in evapotranspiration.

$$\text{WUE (kg ha-mm}^{-1}\text{)} = \frac{Y}{ET^c}$$

where, WUE = Water use efficiency (kg ha-mm^{-1}).

Y = Marketable crop yield (kg ha^{-1}).

ET_C = Crop evapotranspiration (mm).

Field water use efficiency: Field water use efficiency is a ratio between marketable crop yield and field water supply which includes water used by the plant in metabolic activities, ET and deep percolation losses.

$$\text{FWUE (kg ha-mm}^{-1}\text{)} = \frac{Y}{WR}$$

where, FWUE = Field water use efficiency (kg ha-mm^{-1}).

Y = Crop yield in kg ha^{-1}.

WR = Water requirement.

Water use efficiency of selected crops is given in Table 3.1.

Table 3.1 : Water use efficiency of selected crops

Crop	Water requirement (mm)	Grain yield (kg ha^{-1})	WUE (kg ha-mm^{-1})
Rice	1200	4500	3.7
Sorghum	500	4500	9.0
Pearlmillet	500	4000	8.0
Maize	625	5000	8.0
Groundnut	506	4616	9.2
Wheat	280	3534	12.6
Fingermillet	310	4137	13.7

Growing marginalisation and fragmentation of land holdings coupled with rising incidence of informal tenancies and poor rural infrastructure such as road, electricity, markets and education affect factor productivity. Availability of good quality irrigation water coupled with flexibility of irrigation and drainage system and appropriate methods of application as well as pricing of irrigation water is crucial for sustainable use of land and water resources.

In one sense, efficiency of a firm is its success in producing as large an amount of output as possible from given sets of inputs. Maximum efficiency of a firm is attained when it becomes impossible to reshuffle a given resource combination without decreasing the total output.

Linear Programming (LP)

The word linear is used to describe the relationship among two or more variables which are directly proportional. For example, doubling (or tripling) the production of a product will exactly double (or triple) the profit and the required resources, then it is linear relationship. Programming implies planning of activities in a manner that achieves some optimal result with restricted resources.

Linear programming was developed by George B Dantzing (1947) during Second World War. It has been widely used to find the optimum resource allocation and enterprise combination.

Linear programming is defined as the optimisation (minimisation or maximisation) of a linear function subject to specific linear inequalities or equalities.

$$\text{Maximise } \sum_{i=1}^{n} C_j X_j$$

$$\text{Such that } \sum_{i=1}^{n} a_{ij} < b_i$$

where,

$Xj = 0$

Cj = Net income from *j*th activity.

Xj = Level of *j*th activity.

a_{ij} = Amount of *i*th resource required for *j*th activity.

b_i = Amount of $_i$th resource available.

Assumptions of Linear Programming

- **Linearity:** It describes the relationship among two or more variables which are directly proportional.
- **Additivity:** Total input required is the sum of the resources used by each activity. Total product is sum of the production from each activity.
- **Divisibility:** Resources can be used in fractional amounts. Similarly, the output can be produced in fractions.
- **Finiteness of activities and resource restrictions:** There is limit to the number of activities and resource constraints.
- **Non negativity:** Resources and activities cannot take negative values. That means the level of activities or resources cannot be less than zero.
- **Single value expectations:** Resource supplies, input-output coefficients and prices are known with certainty.

Advantages of LP

- Allocation problems are solved.
- Provides possible and practical solutions.
- Improves the quality of decisions.
- Highlights the constraints in the production.
- Helps in optimum use of resources.
- Provides information on marginal value products (shadow prices).

Limitations of LP

- Linearity.
- Considers only one objective for optimisation.

- Does not consider the effect of time and uncertainty.
- No guarantee of integer solutions.
- Single valued expectations.

3.5 Resource Re-cycling

Organic wastes from crops and livestock production and household wastes are the valuable sources for cycling organic residues.

3.5.1. Cycling Organic Wastes

Re-cycling of organic wastes such as crop residues, dung and urine from domesticated animals and wastage from slaughter house, human excreta and sewage, bio-mass of weeds, organic wastes from fruit and vegetables production and household wastes, sugarcane trash, oilcakes, press mud and fly ash from thermal power plants leads to "gold from garbage". Material not suitable for direct application can better apply by **composting** and **vermicompost.**

The ultimate goal of **sustainable agriculture** is to develop farming system that are productive and profitable, conserve the natural resources base, protect the environment and enhance health and safety and to do so over the long term. Two farming systems have been proposed for ensuring sustainability: **low input sustainable agriculture (LISA)** and **organic farming.**

Low Input Sustainable Agriculture (LISA)

In this system, use of external production inputs is minimum. Production costs are obviously lower. Overall risk of the farmer is considerably reduced. Besides, these advantages, pollution of surface and groundwater is avoided and healthy food with very little or no pesticide residues are ensured. These systems held promise for both short and long-term profitability. However, the system suffers from one serious drawback- continuation of low external input agriculture will perpetuate to the vicious circle of low inputs low yields, which the third world countries with ever increasing population pressure can ill-offered.

High input system, on the other hand, will fail sooner or later, as they are not economically and environmentally sustainable. What is the solution then? The optimal input farming system has the premise of low input per unit of output and lays emphasis on **law of diminishing returns.**

3.5.2. Organic Farming and Waste Re-cycling

Organic farming is the backbone of **sustainable agriculture.** Organic farming mainly depends on **organic re-cycling.** Industrial agricultural chemicals like fertiliser, pesticide, herbicide etc are not used or used to minimum extent necessary in this kind of farming. Organic farming may result in compatable performance as **conventional agriculture** and crops growth with high organic manure application could tolerate the pest and disease attack better. It is sound and sustainable way of growing more food.

Organic re-cyclable waste include—crop residues, waste, farm industrial waste, multiple and sewage wastes. They are valuable sources of plant nutrient and humus in tropical and sub-tropical soils of India, where there is general deficiency of organic carbon and plant nutrients due to rapid loss of this component by **bio-degradation.** To make up for these losses, extensive utilisation of organic residues in agriculture is essential. In addition they also protect the soil from erosion.

In India, there is great potential for utilisation of crop residues/ straw of major crops. About 141.2 M t of available straws in the country contributes to about 0.7, 0.84 and 2.1 M t N, P and K respectively. Even if 50 per cent of crop residue is used as animal feed, the rest can be re-cycled. Wide **C:N ratio** of the crop residue leads to immobilisation of plant nutrients. Composing the residues with efficient microbial inoculants is a must to bring down the C:N ratio.

3.6 Flow of Energy in Different Farming Systems

Natural and **agricultural ecosystems** of which humans are a part are fundamentally a network of energy and mineral flows. Green plants capture solar energy and convert it into chemical energy for use by the biological system using the elements of C, **H**, **O**, N, P, K, Ca, Mg and others. Basic to the survival of humans and other animals are adequate suppliers of food that are supplied by the ecosystem.

In agriculture, energy is important in terms of crop production and agro-processing for value adding. Since efficient use of the energy resources is vital in terms of increasing production, productivity, competitiveness of agriculture as well as sustainability of rural living, energy auditing is one of the most common approaches to examining energy efficiency and environmental impact of the production system. It enables researchers to calculate output- input ratio, relevant indicators and energy use patterns in an agricultural activity. Also, the energy audit provides sufficient data to establish functional forms to investigate the relationship between energy inputs and outputs. Estimating these functional forms is very useful for determining elasticity of inputs on yield and production.

Energy requirements in agriculture are divided into two groups – direct and indirect. **Direct energy** is required to perform various tasks related to crop production processes, such as land preparation, irrigation, interculture, threshing, harvesting and transportation of agricultural inputs and farm produce. **Indirect energy,** on the other hand, consists of the energy used in the manufacture, packaging and transport of fertilisers, pesticides and farm machinery. As the term addresses, indirect energy is not directly used on the farm. Major items for indirect energy are fertilisers, seeds, machinery production and pesticides. Calculating energy input in agricultural production is more difficult in comparison to the industry sector due to the high number of factors affecting agricultural production.

Success of agricultural production is measured by the amount of biomass energy captured in the crop as a result of manipulating plants, land and water, while using human and animal power and fossil energy power.

Solar energy reaching a hectare during the year in temperate North America averages about 14 x 10^9 kcal. During a 4 months summer growing season in the temperate region, nearly 7 x 10^9 kcal reach an agricultural hectare. Under favourable conditions of moisture and soil nutrients, com is considered one of the most productive food and feed crops per unit area of land. For example, high yielding maize crop grown on good soils of Iowa can produce about 7,000 kg ha^{-1} of maize grain plus another 7,000 kg ha^{-1} of biomass as stover. Converted to heat energy, this totals 63 x 10^6 kcal (heat energy = 4,500 kcal kg^{-1}) and represents about 5 per cent of the solar energy reaching the hectare during the year (1% during the growing season).

For other crops, the efficiency conversion is much less than for com. For example, potatoes with a yield of **40,000** kg ha^{-1} have a dry weight of about **8,000** kg ha^{-1}. Based on total biomass produced of **12,000** kg ha^{-1} and an energy value of **54 x 10^6** kcal, potatoes have a **0.4** per cent efficiency of conversion or a wheat crop, yielding **2,700** kg ha^{-1} of grain, produces a total of **6,750** kg biomass ha^{-1}, which has a heat energy value of **30 x 10^6** kcal. The **conversion efficiency** of sunlight into biomass in this system is only **0.2** per cent. All of these systems are relatively efficient, however, **agricultural ecosystem** as a whole including pastures and rangeland are less efficient and the average rate of conversion is about 0.1 per cent.

When a natural system capable of producing certain amount of energy containing biomass is converted into an agro-ecological system, the natural capability limit is often exceeded by adding energy inputs. The greater the input of external energy, the more the natural capability of the system can be exceeded and the less sustainable the system becomes. Because of this relationship, an analysis of agro-ecosystem's input/output energy balance can be a comprehensive indicator of its sustainability. In this regard, efficient use of energy by the agriculture sector seems as one of the conditions for sustainable agriculture because it allows financial savings, fossil resources preservation and **air pollution** decrease.

3.6.1 Energy Coefficients

Energy coefficients used in energy calculation are given in Table 3.2.

Table 3.2 : Energy coefficient used in energy calculation (Alipour *et al.*, 2014).

Energy source	Energy coefficient (MJ unit^{-1})	Reference
Human labour (h)	1.96 MJ h^{-1}	Gundogmus (2006)
Fertiliser (kg)		Gundogmus (2006)
N	60.60 MJ kg^{-1}	
P	11.10 kg^{-1}	
K	6.70 kg^{-1}	
O	0.30 kg^{-1}	
Pesticides (kg)		Gundogmus (2006)
Insecticide	199 MJ kg^{-1}	
Fungicide	92 MJ kg^{-1}	
Herbicide	238 MJ kg^{-1}	
Diesel (1) 56.31 MJ l^{-1}		Gundogmus (2006)
Paddy (kg)		Iqubal (2007)
Seed	14.57 MJ kg^{-1}	
Straw	12.50 MJ kg^{-1}	
Machinery (h)	62.70	Gundogmus (2006)
Water (m^3)	0.63	Gundogmus (2006)
Electricity (kWh)	11.93	Gundogmus (2006)

3.6.2 Energy Indices

Energy use efficiency (energy ratio), energy productivity, water productivity, specific energy and net energy can be calculated using the following equations, as suggested in the literature :

$$\text{Energy efficiency} = \frac{\text{Total energy output (MJ ha}^{-1}\text{)}}{\text{Total energy input (MJ ha}^{-1}\text{)}}$$

$$\text{Energy productivity} = \frac{\text{Gain yield (kg ha}^{-1}\text{)}}{\text{Total energy input (MJ ha}^{-1}\text{)}}$$

$$\text{Specific energy} = \frac{\text{Total energy input (MJ ha}^{-1}\text{)}}{\text{Gain yield (kg ha}^{-1}\text{)}}$$

$$\text{Water productivity} = \frac{\text{Gain yield (kg ha}^{-1}\text{)}}{\text{Amount of water applied (m}^3\text{ha}^{-1}\text{)}}$$

$$\text{Net energy} = \frac{\text{Energy output (MJ ha}^{-1}\text{)}}{\text{Energy input (MJ ha}^{-1}\text{)}}$$

Amounts of inputs used and output in rice production for each item are illustrated in Table 3.3. As it can be seen, 137.40 kg nitrogen, 77.1 kg phosphorus, 40.78 kg potassium, 10.63 kg manure, 32.73 1 diesel fuel, 4.41 kg pesticide, 29,345.1 m^3 water, 419.65 h human labour, 24.88 h machinery, 1,112.1 kwh electrical energy per hectare are employed for the production of rice. Average annual yields including seed and straw were found to be 3,033.7 and 3,574.32 kg ha^{-1} in the enterprises that are analysed.

Total energy used in various farm operations was 47,603.82 MJ ha^{-1} (Table 3.3) that consists of 1.87 per cent pesticide (the share of insecticides, herbicides and fungicides of total energy were 0.847, 0.91 and 0.11 per cent, respectively), 1.72 per cent human labour, 3.28 per cent machinery, 19.88 per cent fertiliser, 3.9 per cent fuel (diesel energy was mainly consumed for land preparation, cultural practices and transportation), 27.87 per cent electricity, 38.84 per cent water and 2.68 per cent seed inputs.

The highest energy inputs are provided by water and electricity. Irrigation operation consumed the maximum energy on rice farm due to the higher water requirement of rice crop. Based on the energy equivalents of the input and output given in Table 3.3, the average total energy consumed per farm per year was determined as 47,603.82 MJ ha^{-1}.

Table 3.3 : Energy inputs, outputs and output-input ratio in rice production.

Energy source	Quantity used per unit area (ha)	Total energy equivalent (MJ ha^{-1})	Percentage
Inputs (units)			
Human labour (h)	419.65	822.53	1.72
Machinery (h)	24.88	1,559.89	3.28
Fertiliser (kg)	–	9,459.13	19.88
N (kg)	137.40	8,327.01	17.50
P(kg)	77.10	855.66	1.80
K (kg)	40.78	273.27	0.573
Manure (kg)	10.63	3.19	0.007
Chemicals (kg)	–	888.90	1.87
Insecticides (kg)	2.02	403.26	0.847
Fungicides (kg)	0.57	52.37	0.11
Herbicides (kg)	1.82	433.27	0.91
Diesel (1)	32.73	1,842.95	3.90
Electricity (kWh)	1,112.10	13,267.17	27.87
Irrigation water (m^3)	29,345.10	18,487.40	38.84
Seeds (kg)	87.57	1,275.85	2.68
Total energy input (MJ ha^{-1})	–	47,603.82	100
Outputs (units)	–	–	–
Seed (kg)	3,033.70	44,201.02	48.75
Straw (kg)	3,574.32	46,479.01	51.25
Total energy output (MJ ha^{-1})	–	90,680.04	100
Net energy (MJ ha^{-1})	–	43,076.22	–
Specific energy (MJ kg^{-1})	–	15.70	–
Energy efficiency	–	2.19	–
Energy productivity (MJ kg^{-1})	–	0.064	–
Water productivity (kg m^{-3})	–	0.11	–

3.6.3 Flow of Energy in Different Farming Systems

Flow of energy in different integrated farming systems has been briefly presented for selected farming systems.

Crop-Livestock Farming System

Diversified systems consist of components, such as crops and livestock that coexist independently from each other. In this case, integrating crops and livestock serves primarily to minimise risk and not to re-cycle resources. In an integrated system, crops and livestock interact to create a synergy, with re-cycling allowing the maximum use of available resources. Crop residues can be used for animal feed, while livestock and livestock by-product production and processing can enhance agricultural productivity by intensifying nutrients that improve **soil fertility,** reducing the use of chemical fertilisers. A high integration of crops and livestock is often considered as a step forward, but small farmers need to have sufficient access to knowledge, assets and inputs to manage this system in a way that is economically and environmentally sustainable over the long term.

Livestock provide dairy products and meet to household besides supplying dung for bio-gas and manure to the crops. Crops provide grains, vegetables and fruits to household. Resource flow model for crop-livestock farming system has been given in Fig 3.3.

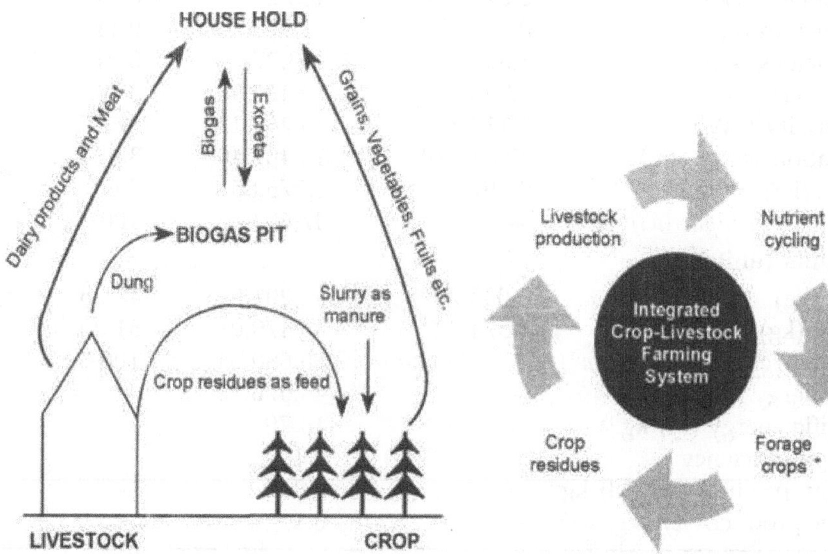

Fig. 3.3 : Resource flow model for crop-livestock farming system.

Major farming systems in this category include:

- Crops-poultry
- Crops-sheep
- Crops-goat
- Crops-dairy animals
- Rice-fish
- Rice-mushroom, etc.

Crop-Livestock-Forestry System

This system combines cropping, livestock and forestry activities to promote the recovery of degraded pastures. Each farm will have a varied production system, such as grains, fibers, meat and milk and agro-energy. It also aims to improve **soil fertility** with the use of adequate **cropping systems** and techniques to optimise and intensify its use. Therefore, it allows the diversification of economic activities on farm and minimises income risks due to climate and/ or market changes. The system consists in growing forest species simultaneously with commercial crops like soybeans, maize or beans added for the first 2 or 3 years. After crop harvest, the area is planted with forages for livestock, associated with maize or sorghum. After grain harvesting, the pasture is already established between the tree rows, enabling grazing, until wood is harvested.

Forestry component supplies feed to the livestock and nutrients to the crops. Livestock supply manure to forestry as well as crops. The crops supply feed to the livestock. Resource flow model for crop-livestock-forestry farming system has been given in Fig 3.4.

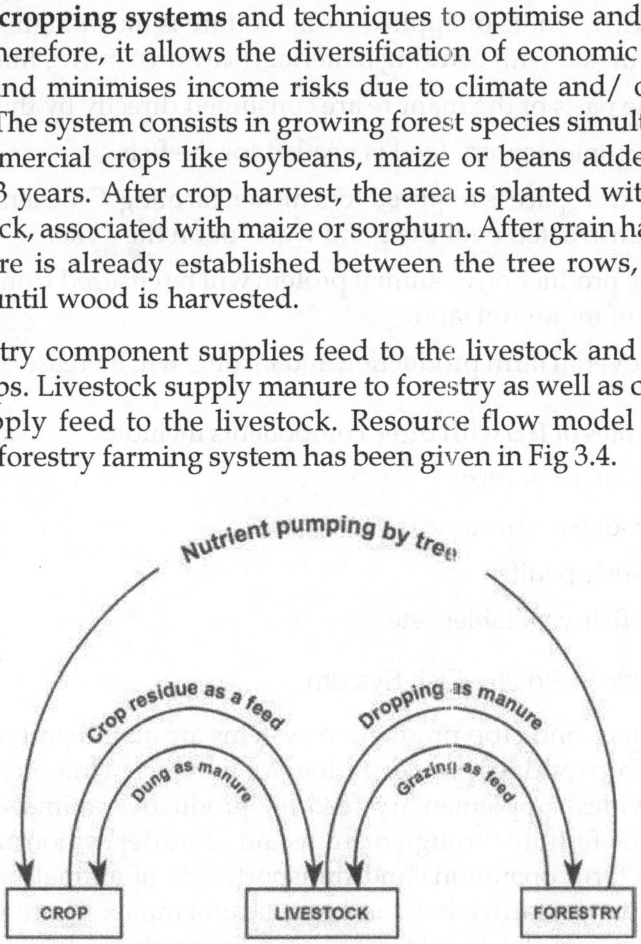

Fig. 3.4 : Resource flow model for crop-livestock-forestry farming system.

Crop-Fish-Poultry System

Following are some of the additional advantages, when crop and fish culture is integrated with chicken rising on/ or near the pond dykes:

- ❐ As usual, food crops (rice, maize, pulse crops, groundnut, etc.) meets the needs of the family.

- Grain and other by-products of the crops meets the needs of poultry feed.
- Direct discharge of fresh and chicken manure to the fish ponds produces enough natural fish feed organisms without the use of any additional manure/fertiliser.
- Nutritive value of applied fresh manure is much higher than dry and mixed with bedding materials (saw dust or rice husk).
- Some parts of the manure are consumed directly by the fish.
- No supplementary feed is needed for the fish.
- No extra space is required for chicken farming. Chicken sheds can be constructed over the pond water or on the dyke.
- More production of animal protein will be ensured from the same area of minimum land.
- The overall farm production and income will increase.

Examples of IFS with three components include:

- Crop-dairy-poultry
- Crop-dairy-agro-forestry
- Rice-fish-poultry
- Rice-fish-vegetables, etc.

Crop-Livestock-Poultry-Fish System

Livestock and crop production systems are an integral part of one another. Crop residues provide fodder for livestock while, occasionally, grain provides supplementary feed for productive animals. Animals improve soil fertility through manure and urine deposition and animal power for farm operations and transport. Sale of animals sometimes provides cash for farm labour and agricultural inputs. There are several examples of completely integrated crop-livestock production systems where sustainable increases in both crop and livestock production have been achieved after considerable periods (30-40 years) of continuous cropping without resulting in land degradation.

Introduction of poultry and fish components in crop-livestock system leads to additional benefits as indicated below:

- Ruminants to utilise nonmarketable crop residues *in situ,* the manure production, from which can be used by fish.

- Availability of water plants which can also be simultaneously utilised by ruminants.
- Reduced cost of feeding and production due to more intensive use of available indigenous feeds.
- Use of purchased concentrate supplements to fish and poultry components is restricted to dairy cattle.
- Demonstrable benefits in terms of significant income generation from the sale of crops, animals, fish and some by-products like rice-bran in rice-based systems.
- With small farms, the meager resources are put to more effective.

Resource flow model for crop-livestock-poultry-fish system (Fig 3.6) indicates efficient use of resources for improving farm productivity, employment generation and minimising the effects of climate change.

The key success to these farming systems is effective components integration involving the recycling of nutrients within the system. A particular challenge facing farmers is to minimise nutrient losses through good management.

Intensive farming system components include:

1. Rice-fish-poultry-livestock
2. Rice-fish-mushroom-poultry
3. Cropping-fish-pigeon-agro-forestry
4. Cropping-fish-poultry-mushroom
5. Cropping-dairy-poultry-agro-forestry, etc.

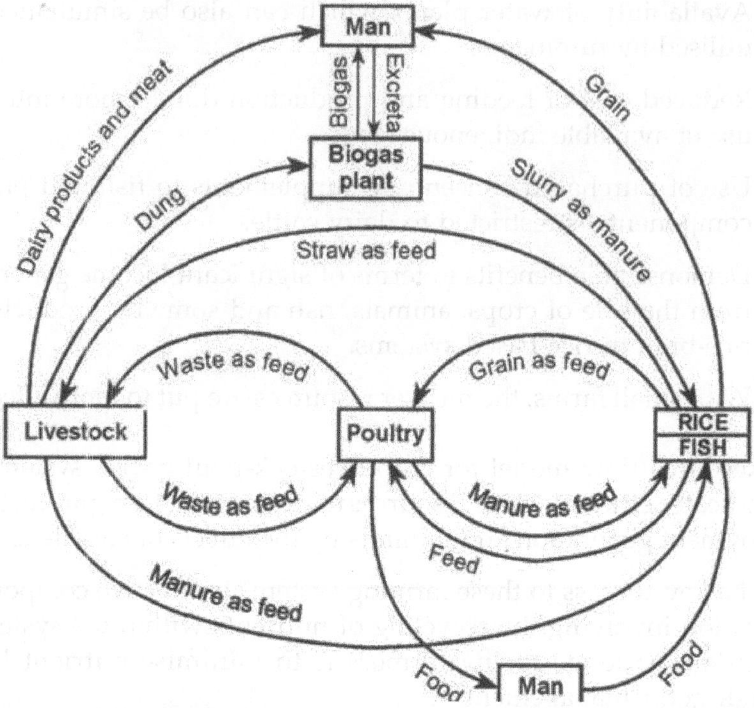

Fig. 3.5 : Resource flow model for crop-livestock-poultry-fish system.

3.7 Farming System and Environment

Food security, livelihood security, water security, natural resource conservation and environmental protection have emerged as major issues worldwide. The IFS was first propagated with overall aim of establishing sustainable farming system with reduction in exposure of the environment to pesticides and other agro-chemicals. Adoption of organic farming was naturally expected to prevent short and long term ill effects on the biosphere. Programmes, such as linking environment and farming assisted farmers to move towards IFS. The IFS is targeted to bring a commitment to good husbandry and animal welfare along with efficient soil management and appropriate cultivation techniques using crop rotations and improved cultivars, reducing use of crop protection chemicals and fertilisers with the aim to maintain the landscape, rural communities and wildlife habitats. It is a dynamic concept with flexibility to adopt new technology and change market pressure and consumer demands and to meet the future expectations through sustainable farming system which cares for the environment.

Table 3.4 : Contribution of sustainable farming system practices to important ecosystem services.

Objective	System component			
	Match cover	**Minised or no tillage**	**Legumes to supply plant nutrients**	**Crop rotation**
Simulate optimum 'forest-floor conditions	*	*		
Reduce evaporative loss of moisture from soil surface	*			
Reduce evaporative loss from upper soil layers	*	*		
Minimise oxidation of soil organic matter and loss of CO_2		*		
Minimise soil compaction	*	*		
Minimise temperature fluctuations as soil surface	*			
Provide regular supply of organic matter as substrate for soil organism activity	*			
Increase, maintain nitrogen levels in root zone	*	*	*	?
Increase catio exchange capacity of root zone	*	*	*	?
Maximise rain infiltration, minimise runoff	*	*		
Minimise soil loss in runoff and wind	*	*		
Permit, maintain natural layering of soil horizons through action of soil biota	*	*		
Minimise weeds	*	*		?
Increase rate of biomass production	*	*	*	?
Speed recuperation of soil porosity by soil biota	*	*	*	?
Reduce labour input		*		
Reduce fuel/energy inputs		*	*	?
Re-cycle nutrients	*	*	*	?
Reduce pest-pressure of pathogens				?
Re-build damaged soil conditions and dynamics	*	*	*	?
Pollination services	*	*	*	?

One of the main requirements for ecologically sustainable production is healthy soil, creating an environment in the root zone that optimises soil biota activity and permits root functioning to the maximum possible extent. Roots are able to capture plant nutrients and water and interact with a range of soil micro-organisms beneficial to soil health and crop performance. Maintenance or improvement of soil organic matter content, soil structure and associated porosity are critical indicators of sustainable production and other ecosystem services. Contribution of selected sustainable farming practices to improve ecosystem services are presented in Table 3.4.

However, now there exists two schools of thoughts, one considering IFS as boon for environment while the other takes it as curse. It appears from the literature that it is the way one takes IFS to perceive its benefits and dangers. Environmental implications of IFS are briefly presented under different heads.

3.7.1. Ecosytem

This category comprises floral and faunal bio-diversity, habitat diversity and landscape conservation. The IFS performs better than **conventional farming** in respect to floral and faunal diversity. Due to the ban of synthetic pesticides and N-fertilisers, organic farming system provides potentials that result in positive effects on wildlife conservation and landscape. Potentially, organic farming leads to a higher diversity of wildlife habitats due to more highly diversified living conditions, which offer a wide range of housing, breeding and nutritional supply.

Integrated fish farming seems to favour antimicrobial-resistant bacteria in the pond environment which is attributed to the selective pressure of antimicrobials there and to the introduction of antimicrobial resistant bacteria from animal manure. This change may pose a potential risk of destabilisation of natural micro-flora in the ecosystem which is ultimately deleterious to the environment.

3.7.2. Soil

The IFS tends to conserve soil fertility and system stability better than conventional farming system. This is due to mostly higher organic matter contents and higher biological activity in organically farmed soils than in conventionally managed. Furthermore, IFS has high erosion control potential.

3.7.3. Ground and Surface Water

The results show that IFS results in lower or similar nitrate leaching rates than **integrated** or **conventional agriculture.** Actual leaching rates per hectare are up to 57 per cent lower on IFS than on conventional fields. **Organic farming** does not pose any risk of ground and surface water pollution from synthetic pesticides.

3.7.4. Climate and Air

On per-hectare scale, the CO_2 emissions are 40-60 per cent lower in IFS than in conventional ones, whereas on per-unit output scale, the CO_2 emissions tend to be higher in organic farming system. The N_2O emissions per hectare on organic farms tend to be lower than on conventional farms. The IFS has a lower CH_4 emission potential on a per hectare scale, while CH_4 emissions per kg of milk are estimated to be higher in organic dairy farms than in conventional ones. Calculations of NH_3 emissions in organic and conventional farming system conclude that organic farming bears a lower NH_3 emission potential than conventional farming system. Due to the fact that synthetic pesticides are not permitted in organic farming, significantly lower air contamination is ensured than in conventional farming.

3.7.5. Farm Input and Output

Nutrient balances of IFS farms in general are close to zero. The N, P and K surpluses of IFS were significantly lower than on conventional farms. Negative balances were found for P and K. Most research studies reviewed indicate that energy consumption on organic farms is lower than on conventional farms.

3.7.6. Animal Health and Welfare

Housing conditions and health status depend highly on farm specific conditions, thus housing conditions seem not to differ significantly between IFS and conventional farms. IFS dairy cows tend to have a longer average productive life than conventional dairy cows.

3.7.7. Quality of Food Produced

The risk of contaminating food with pesticides and nitrate can be assumed to be lower in IFS rather than in conventionally produced food. A strong argument exists for the superiority of animal products from organic in comparison to conventional farming.

3.7.8. IFS Associated Health Risk

Increasing health consciousness among the people has led to the increased demand for vegetable products grown in the absence of pesticides and synthetic fertilisers, popularly known as **organic foods.** Moreover, land available for food production is decreasing day-by-day. As such, **intensive cropping** system and intensive livestock system have come to merge into IFS to sustain the ever growing demand for food and feed. For realising higher yield, farmers augment **soil fertility** through application of **compost** and irrigation with **sewage** in most of the vegetable growing regions in India. Sewage and compost, being common source of **zoonotic pathogens,** may contaminate soil with the pathogens and the transmission of infection through vegetable produce grown on contaminated soil is not rare. Studies on *Salmonella,* a zoonotic pathogen, in plants like cowpea, greengram, vegetables and maize reveal that the pathogen rapidly spread to all parts of the plants within few hours of crop irrigation with contaminated water. The possibility of circulation of other sewage bacteria and potential for opportunistic bacteria becomes more apt in IFS model because of closely knit livestock, fishery and human components (Singh 2011).

The IFS may result in origin of new pathogens or re-assortment of older ones into new forms. Influenza pandemic possibility possibly originated from pig-duck system, facilitating, re-assortment of avian and mammalian influenza virus. Cysticercosis and Cryptosporidium infection may also be transmitted through contaminated and salad vegetables irrigated with contaminated water. It is well-known that pigs and other animals under such IFS may become carriers of methicillion-resistant *Staphylococcus aureus* **(MRSA)** and other dangerous pathogens due to many reasons, posing a serious health threat to the community.

In the recent past, anthropogenic changes, largely in land use and agriculture, particularly IFS, are implicated in the apparent increased frequency of emergence and re- emergence of zoonoses and multiple drug resistant **(MDR)** infections. Excreta from poultry or livestock suffering from some zoonotic infectious disease may be transmitted to the fish and vegetables which constitute important components of human diet.

References

Behera UK and France J 2016. Integrated farming system and livelihood for small and marginal farmers in India and other developing countries. Advances in Agronomy, Vol 138: 235-274.

Bhan S and Behera UK 2014. Conservation agriculture in India—Problems, prospects and policy issues. International soil and water conservation research, Vol 2(4): 1-12.

CARDI 2010. A Manual on Integrated Farming System (IFS). Caribbean Agricultural Research and Development Institute.

FAO 2006. Conservation agriculture—Advantages and disadvantages, http://www.fao.org/ag/ca/lc.html.

FAO 2015. What is conservation agriculture? http://www.fao.org/ag/ca/la.html.

Gliesmann SR 2004. The agro-ecosystem concept - University of Colorado Boulder. Department of Environmental Studies. University of California, Santa Cruz, CA. www.colorado.edu/eeb/.../ Gliesmann Ch 2. The agro-ecosystem concept.pdf.

Gundogmus E 2006. Energy use on organic farming: a comparative analysis on organic versus conventional apricot production on small holdings in Turkey. Energy Conversation Management, 47: 3351-3359.

Jayanthi C, Devasenapathy P and Vennila C 2008. Farming system. Satish Serial Publishing House, Azadpur, New Delhi, India.

MAFF 2000. Towards sustainable agriculture—A pilot set of indicators. Ministry of Agriculture Fisheries and Food, www.adlib.ac.uk/resources/000/015/650/pilotindicators.pdf.

Mary V Gold 2007. Sustainable agriculture: Related terms. Alternative Farming System Information Center, National Agricultural Library, US Department of Agriculture, 10301 Baltimore Avenue, Beltsville, MD 20705-2351.

NMSA 2010. National Mission for Sustainable Agriculture: Strategies for meeting the challenges of climate change. Department of Agriculture and Co-operation, Ministry of Agriculture, New Delhi.

Pimentel D 1985. Energy flows in agriculture and natural ecosystem. Department of Entomology and Section of Ecology and Systematics, Cornell University, Ithaca, NY 148.53. www.utdallas.edu/ -mjleaf/PimentalEnerg.

Rajju PS, Mittu K and Rajesh L 2014. Integrated farming system. IOSR Journal of Agriculture and Veterinary Science, Vol 7 (10): 36-42.

Rambodagedera DB 2008. Sustainable agriculture. National Science Foundation, thakshana.nsf.ac.lk/ slstic/NA-354/NA-354.pdf.

Rana SS 2015. Recent advances in integrated farming system - Himachal Pradesh, hillagric.ac.in/edu/coa/..7607.pdf.

Rana SS and Pankaj Chopra 2013. Integrated farming system. Department of Agronomy, College of Agriculture, CSK Himachal Pradesh Krishi Vishvavidyalaya, Palampur.

Reddy SR 2016. Principles of agronomy. Kalyani Publishers, Ludhiana, India.

Reddy SR and Nagamani C 2016. Principles of crop production. Kalyani Publishers, Ludhiana, India.

Singh BR 2011. Environmental risk from integrated farming system. Centre for Animal Disease Research and Diagnosis, Indian Veterinary Research Institute Izatnagar-243122. https://www.researchgate.net/publication/260433503.

Singh RP 2013. Farming system: Concept-scope, www.slideshare.net/rpratapll/farmingsyste.

Subba Rao A and Mandal KG 2007. Modem concepts of agriculture – Sustainable agriculture. Indian Institute of Soil Science, Nabibagh, Berasia Road, Bhopal-462 038, India, nsdl.niscair.res.in/jspui/bitstream.

❑❑❑